Nanoparticles and Nanodevices in Biological Applications

Lecture Notes in Nanoscale Science and Technology

Volume 4

Series Editors

Volumes Published in This Series:
Volume 1: Self-Assembled Quantum Dots, Wang, Z.M., 2007
Volume 2: Nanoscale Phenomena: Basic Science to Device Applications, Tang, Z., and Sheng, P., 2007
Volume 3: One-Dimensional Nanostructures, Wang, Z.M., 2008
Volume 4: Nanoparticles and Nanodevices in Biological Applications, Bellucci, S. (Ed.), 2009

Forthcoming Titles:
B-C-N Nanotubes and Related Nanostructures, Yap, Y.K., 2008
Towards Functional Nanomaterials, Wang, Z.M., 2008
Epitaxial Semiconductor Nanostructures, Wang, Z.M., and Salamo, G., 2009

Stefano Bellucci (Ed.)

Nanoparticles and Nanodevices in Biological Applications

The INFN Lectures - Vol I

Stefano Bellucci
Istituto Nazionale di Fisica Nucleare
Laboratori Nazionali di Frascati
via E. Fermi, 40
00044 Frascati RM
Italy
bellucci@lnf.infn.it

ISBN: 978-3-540-70943-5 e-ISBN: 978-3-540-70946-6

Library of Congress Control Number: 2008933215

Printed on acid-free paper

9 8 7 6 5 4 3 2 1

springer.com

Preface

This is the first volume in a series of books on selected topics in nanoscale science and technology, based on lectures given at the well-known Italian Institute for Nuclear Physics (Instituto Nazionale di Fisica Nucleare, or in short INFN) schools of the same name (Nanoscience & Nanotechnology 2006, Nanosceience & Nanotechnology 2007). The aim of this collection is to provide a reference corpus of suitable, introductory material to relevant subfields, as they mature over time, by gathering the significantly expanded and edited versions of tutorial lectures, given over the years by internationally known experts.

The field of nanoscience has witnessed a rapid growth in the last decade. Recently, the attention of the community of nanoscientists has been focusing more and more on technological applications. Nanotechnology is an enabling technology, with a high potential impact on virtually all fields of human activity (industrial, health-related, biomedical, environmental, economy, politics, etc.). Its potential yields high expectations for solutions to the main needs of society, although it issues of sustainability and compatibility need to be addressed. The fields of research application in nanoscience include aerospace, defence, national security, electronics, biology, and medicine.

A strong interest in assessing the current state of the art of the fast-growing fields of nanoscience and nanotechnology, as well as the need of stimulating research collaboration, prompted me to promote and direct the International School and workshop, "Nanoscience & Nanotechnology (n&n 2006)" on November 6–9, 2006 under the patronage of INFN, the University of Rome Tor Vergata, and the Catholic University of Rome, with generous sponsorship by 3M, 2M Strumenti, Physik Instrumente, and RS Components. The aims of this event were manifold:

- To foster the concrete planning of future devices based on innovative (nano) materials, involving both industrial entities and public research institutes
- To allow the presentation by sponsoring firms of their instrumentation and success stories based on current use by significant customers
- To lend an opportunity for preparing and presenting joint projects, involving both industry and public research (see e.g., the EU Framework Programs)
- To explore the possibility of integrating nanodevices from their concept into system projects

In this context, tutorial lectures were delivered at the school, addressing general and basic questions about nanotechnology, such as what it is, how does one go about it, and what purposes can it serve. In these tutorial sessions, the nature of nanotechnology, the instruments of current use in its characterizations, and the possible applicative uses were described at an introductory level. Given the great success and broad range of these lectures, it was decided to publish them over time as a collection of well-edited topical volumes.

The present set of notes results, in particular, from the participation and dedication of many prestigious lecturers and colleagues. As usual, the lectures were carefully edited and reworked, taking into account the extensive follow-up discussions.

A tutorial lecture by *Vincenzo Balzani* and collaborators (Univ. Bologna, Italy) introduces the reader to the topic of molecular devices and machines, seen as a journey into the nano world. The *Santina Carnazza* (Univ. of Messina, Italy) contribution deals with surface biofunctionalization (by controlled ion implantation and fibronectin adsorption) aimed to enhance promonocytic cell adhesion and spatial confinement, and micropatterning of polymer surfaces (by controlled ion irradiation on stripes of given dimensions) to obtain alignment and controlled positioning of adherent fibroblasts. The former may be important for biosensing, and the latter in preparing cell-based integrated circuits; hence having an impact both in biomedicine, particularly in regenerative medicine (including tissue engineering), and in BioMEMS applications. *Andrea Salis* and coworkers (Univ. Cagliari, Italy) go about biotechnological applications of lipases immobilized onto porous materials—i.e., biodiesel production and biosensors. For biosensing use, the immobilization of the lipases was performed by the Cagliari group on porous silicon.

Cellular interactions with engineered nanoparticles are dependent on many variables—some inherent to the nanoparticle (size, shape, surface reactivity, degradation, agglomeration/dispersal, and charge), and some due to the inherent properties of the cells or tissues responding to the nanoparticle (cell type; cell surface interactions with the nanoparticle; whether cellular membranes have pores that allow or block passage of nanoparticles, cellular enzyme degradation of the outer protective surface revealing a toxic nanoparticle core; cellular storage of nanoparticles or degradation products (bioaccumulation), within the cell ultimately causing the cell's death). Functionalization and shorter exposure times increased biocompatibility; however, nanoparticle size and reactivity in relation to the type of cells and organs to be targeted seemed to be equally important. Understanding the biological effects of nanoparticles at the gross (microPET) and microscopic levels (light and electron microscopy) is essential in predicting nanoparticle processing, degradation and excretion in cells, and mammalian systems in general. In this respect, in her lecture, *Barbara Panessa-Warren (BNL, USA)* provides the reader with an overview of the types of phenomena that have been reported in the literature with living cells and tissues exposed to nanoparticles, as well as new experimental data on the biological cell and tissue responses in vitro (using human lung and colon epithelial monolayers) and in vivo (in mice) to nanoparticles designed for biomedical use (prepared with and without surface functionalization); with specific attention directed to how dose, exposure time, and surface reactivity affect biocompatibility and cytotoxicity. A contribution that reviews recent results about the toxicity of nanomaterials, concentrating in particular on carbon nanotubes, is the subject of the tutorial by *Stefano Bellucci (INFN-LNF, Italy)*.

I wish to thank all lecturers, and especially those who contributed to the first volume in this series, for the time and effort put into this book project. I am confident that this first set of lecture, will, in turn, provide an opportunity for those who are just now beginning to get involved with nanoscience and nanotechnology, allowing them to get contacts and prime, up-to-date information from the experts. I wish to especially thank Mrs. Silvia Colasanti for precious help in carrying out organizational and secretarial work.

Total love and gratitude go to my wife Gloria, and our fantastic daughters Costanza, Eleonora, Annalisa, and Erica for providing me relentlessly with endurance, energy, enthusiasm, and patience.

Frascati,
April 2008

Stefano Bellucci

Contents

Nanoparticle Interactions with Living Systems: In Vivo and In Vitro Biocompatibility 1
Barbara J. Panessa-Warren, John B. Warrren, Mathew M. Maye and Wynne Schiffer

Carbon Nanotubes Toxicity 47
Stefano Bellucci

New Advances in Cell Adhesion Technology 69
Santina Carnazza

Light-powered Molecular Devices and Machines 131
Vincenzo Balzani, Giacomo Bergamini and Paola Ceroni

Hofmeister Effects in Enzymatic Activity, Colloid Stability and pH Measurements: Ion-Dependent Specificity of Intermolecular Forces 159
Andrea Salis, Maura Monduzzi and Barry W. Ninham

Index 195

List of Contributors

Vincenzo Balzani
Dipartimento di Chimica "G. Ciamician", Università di Bologna, via Selmi 2,
40126 Bologna, Italy,
vincenzo.balzani@unibo.it

Stefano Bellucci
INFN-Laboratori Nazionali di Frascati, Via E. Fermi 40, 00044 Frascati, Italy,
bellucci@lnf.infn.it

Giacomo Bergamini
Dipartimento di Chimica "G. Ciamician", Università di Bologna, via Selmi 2,
40126 Bologna, Italy

Mathew M. Maye
Center for Functional Nanomaterials Brookhaven National Laboratory Upton, NY
USA 11973–5000

Santina Carnazza
Department of Microbiological Genetic and Molecular Sciences, University of
Messina, Sal. Sperone 31, Vill. S. Agata, 98166 Messina, Italy,
santina.carnazza@unime.it

Paola Ceroni
Dipartimento di Chimica "G. Ciamician", Università di Bologna, via Selmi 2,
40126 Bologna, Italy

Maura Monduzzi
Dipartimento di Scienze Chimiche, Università di Cagliari – CSGI, Cittadella
Monserrato, S.S. 554 Bivio Sestu, 09042 Monserrato, Italy

Barry W. Ninham
Dipartimento di Scienze Chimiche, Università di Cagliari – CSGI, Cittadella
Monserrato, S.S. 554 Bivio Sestu, 09042 Monserrato, Italy; Department of Applied
Mathematics, A.N.U. Canberra, Australia

Barbara J. Panessa-Warren
Department of Energy Sciences and Technology, Brookhaven National Laboratory,
Upton, NY 11973, USA

Andrea Salis
Dipartimento di Scienze Chimiche, Università di Cagliari – CSGI, Cittadella Monserrato, S.S. 554 Bivio Sestu, 09042 Monserrato, Italy

Wynne Schiffer
Medical Department Brookhaven National Laboratory Upton, NY USA 11973–5000

John B. Warren
Instrumentation Division, and Center for Funcional Nanomaterials Brookhaven National Laboratory Upton, NY, USA 11973–5000

Nanoparticle Interactions with Living Systems: In Vivo and In Vitro Biocompatibility

Barbara J. Panessa-Warren, John B. Warrren, Mathew M. Maye
and Wynne Schiffer

1 Introduction

Multiple factors must be considered to predict how specific nanoparticles will react with living biological material. In the current literature, a great deal of attention has been focused on the role of nanoparticle structure in determining biocompatibility and predicting cellular responses. This lecture presents some of the basic concepts and data demonstrating how nanoparticle interactions with biological cells and organisms require an understanding, not only of the physicochemical characteristics and reactive-potential of the nanoparticle, but also knowledge about biological processes and how different cells may respond to specific nanoparticle characteristics. Biological processes and cell structure must be considered when attempting to interpret viability data following nanoparticle exposure, and incorporation and cellular transport of nanoparticles in vivo and in vitro, in order to correctly interpret biocompatibility. Such factors as cell type (blood macrophages, epithelial cells, connective tissue cells, neurons, histiocytes, muscle cells, etc.), cell and tissue function, cell age (neonatal, aged and dedifferentiated cells respond very differently), and the interactions of nanoparticles with cells/tissues lining different physiologic compartments through which the nanoparticle must pass to arrive at the target organ or cells (interactions with endothelial cells lining blood vessels, passing through slits and ducts in the kidney, and pores and fenestrations in capillaries, etc.) must all be considered and factored into the interpretation of the results. For an engineered nanoparticle to

Barbara J. Panessa-Warren
Department of Energy Sciences and Technology, Brookhaven National Laboratory,
Upton, NY 11973, USA

John B. Warren
Division of Instrumentation, and Center for Functional Nanomaterials Brookhaven National Laboratory Upton, NY, USA 11973-5000

Mathew M. Maye
Center for Functional Nanomaterials Brookhaven National Laboratory Upton, NY USA 11973–5000

Wynne Schiffer
Medical Department Brookhaven National Laboratory Upton, NY USA 11973–5000

S. Bellucci (ed.), *Nanoparticles and Nanodevices in Biological Applications*. Lecture Notes in Nanoscale Science and Technology 7,

be used successfully in a biomedical application for diagnostic imaging, drug delivery or facilitated radiotherapy, specific factors must be considered that are unique to working with living cells, such as the:

(a) Characteristics of the targeted cells;
(b) Types of cells and tissues that the engineered nanoparticles will encounter on the way to the targeted cells, and how they will interact with the nanoparticles at each site;
(c) Loss of the calculated nanoparticle dose to the targeted cells, due to nanoparticle (i) engulfment by blood macrophages, (ii) accumulation and storage by tissue histiocytes (macrophage-type cells localized within specific tissues), or (iii) binding with antibodies and/or complement protein triggering aggregation and attachment of the engineered nanoparticles to blood vessel walls and fibrin clots; and
(d) Alterations in the very physicochemical characteristics of the nanoparticle surface in vivo and in vitro due to adsorption, wrapping or entanglement in carbohydrates, proteins or DNA (including biomolecules like serum proteins, different cell surface mucopolysaccharides, proteoglycans), or intracellular attack from lysosomal enzymes and micro-environments at low pH (as in phagocytic vacuoles and during lysosomal degradation of endosomal vesicles).

Sometimes engineered nanoparticles perform beautifully with the proposed target cells, but may interact with blood cells, blood proteins, phagocytic tissue bound cells (histiocytes)which are part of the reticuloendothelial system (RES) found in the liver (Kupffer cells), lung (lung macrophages), skin (Langerhans cells), brain (microglia), and the macrophage type cells within the spleen, thymus and lymph nodes [1]. Nanoparticles often become entangled or coated with mucus produced by epithelial cells, or extracellular matrix structures on the apical surfaces of cells adjacent to the target cells, reducing the number of nanoparticles that could be localized at the desired site. These interactions with phagocytic cells are facilitated by nanoparticle adsorption or wrapping with biological elaborated materials, as well as nanoparticle agglomeration, which in turn initiates:

- Phagocytic attack and engulfment by cells of the reticuloendothelial system (RES) or blood macrophages following antibody/complement binding (opsonization).
- In the case of biomaterial-coated nanoparticles bound to cell surfaces via receptor proteins, the nanoparticles can be transported into the cells by endocytosis, which can cause degradation of the nanoparticle, and possible cellular bioaccumulation with or without possible cytotoxicity.
- Or, complement fixation of the nanoparticles can lead to activation of cellular inflammatory responses, interference in the blood clotting cascade and subsequent fibrin formation and anaphylaxis [2]

This lecture surveys some of the documented responses in the current scientific literature and original data of living biological systems following exposure to different types of nanoparticles, and how both biological factors, and nanoparticle compositional characteristics, as well as dose, exposure time and nanoparticle surface reactivity & functionalization collectively impact biocompatibility [3–5].

2 Biocompatibility Depends on Nanoparticle Characteristics and Biological Factors

In general, cellular interactions with engineered nanoparticles are strongly dependent on variables inherent to the engineered nanoparticles. Such factors as size, morphology, surface charge & reactivity, defects in the nanoparticle surface, elemental composition (core composition), and surface functionalization are the most obvious nanoparticle factors that directly impact biocompatibility.

2.1 In Vivo Evidence

Once in contact with biological cells and tissues, nanoparticles have the ability to further interact with biomolecules producing nanoparticle aggregates, or agglomerations of nanoparticles with proteins, components of the extracellular matrix (ECM), carbohydrates, salts and fragments of the cells themselves (pieces of membrane or cytoplasm). This in turn alters the characteristics of the engineered nanoparticle creating a virtually new ‘nanoparticle-biological material’. These bio-altered nanoparticles wrapped or enveloped with biological substances such as glycoproteins, complement proteins, other serum proteins, glycocalyx material, cellular components etc…) are capable of producing responses quite different from the intended function (of the engineered nanoparticle) in organisms, as well as in tissue culture in vitro systems. Nels et al. [3] stated that the biological and biokinetic properties of nanoparticles such as chemical composition, size, surface structure, solubility, shape and aggregation directly affect biological toxicity through interactions such as protein binding, cell uptake, and transport from the portal of entry to a target site. The binding of nanoparticles in vivo to host proteins (serum proteins, glycoproteins of the extracellular matrix, receptor proteins expressed on the cell surface, etc.), can make the nanoparticles aggregate triggering activation of the complement system resulting in phagocytic cell capture of the aggregates, as well as adherence of aggregated material to blood vessel walls followed by platelet and thrombin activation which could lead to tissue injury due to clot formation and inflammatory damage, as well as anaphylaxis and death [2]. Proteins coming in contact with the highly reactive nanoparticle surface could also undergo denaturation, inducing structural changes in the proteins [6], which may activate phagocytic attack. Salvador-Morales et al. [7] demonstrated that carbon nanotubes activated human complement and bound Clq. Although many different proteins are found in plasma, the nanotubes selectively bound fibrinogen and apolipoproteins in the greatest quantity. Similarly, P. Cherukuri et al. [8] observed that blood proteins displaced surfactant coatings on single-walled carbon nanotubes (SWCNTs) within seconds in the blood. These serum protein-coated SWCNTs at low doses produced no acute signs of toxicity in adult rabbits behaviorally, nor were there signs of cytotoxicity or acute inflammation seen in tissue sections following necropsy. Therefore, surface binding of biological proteins to nanoparticles in some cases can facilitate nanoparticle biocompatibility for a specified time period, yet in other situations, this surface binding of blood proteins such as antibodies and complement can initiate acute and chronic inflammatory processes in vivo. Only experimental testing of each type of engineered nanoparticle with an appropriate in vivo biological model can reveal what modifications of the nanoparticle will improve and facilitate safe and optimized biological implementation. This tutorial will present an overview of the biological responses to specific nanoparticles and how these factors can affect the interpretation of nanoparticle biocompatibility/toxicity results.

2.2 In Vitro Evidence

Some of the same nanoparticle/bio phenomena are found when nanoparticles are incubated with tissue culture cells. Although there is no circulatory system nor complex immune and inflammatory systems to respond to the presence of the nanoparticles in vitro, nanoparticles are capable of binding to cell surfaces, ECM glycoproteins, serum proteins in the culture media [4,7,9]. Cells exposed to nanoparticles in vitro have been used to measure the changes in the up and down regulation of specific cellular proteins (fibronectin, E-cadherin, specific enzymes), changes in cell cycle and the appearance of inflammatory and apoptotic markers [3, 10–13]. Tissue culture cells can have extensive extracellular matrices in culture composed of simple glycoproteins, DNA, proteoglycans and various cellular proteins that are expressed at the apical cell surface in response to the presence of foreign material. Nanoparticles can become bound to these cell-elaborated materials at the cell surface, and subsequently bind to cell surface receptors that facilitate entry of the nanomaterial into the cell (via **endocytosis**). This can result in the lysosomal digestion and processing of the nanoparticle within the cell with recycling of the adsorbed protein and the attached fragments of the plasma membrane (back to the cell), and the eventual expulsion from the cell of the intact or degraded nanoparticle [4, 5]. Nanoparticles may also remain stored within phagocytic vacuoles or endosomal vesicles for indeterminate periods of time [14]. Endocytosis by epithelial cells, fibroblasts and neuronal cells should not be confused with **phagocytosis** by specialized cells called macrophages. Macrophages are found within the blood stream and serve to identify and rid the blood of foreign microbial and non-living materials. There are macrophages associated with specific tissues (lung, brain, liver, skin, spleen, thymus, lymph nodes, etc.) called histiocytes mentioned earlier in this lecture, that specifically function to rid those tissues of debris, dead tissue, microorganisms and particulates by engulfing them into phagocytic vacuoles. Phagocytic cells attempt to digest and degrade engulfed materials within the phagocytic vacuole;

however, when this fails, the phagocyte will retain the material within the vacuole preventing it from reentering the blood or tissue. Both endocytosis and phagocytosis can produce significant amounts of cell sequestered nanoparticles resulting in nanoparticle **bioaccumulation** all of which may or may not cause the cells harm. In the case of quantum dots sequestered for long periods of time within phagocytic vacuoles or endocytic vesicles, there is a risk of the protective capping material (such as ZnS or PEG) deteriorating with time and allowing the release of Cd or Se ions directly into the cells, both of which are toxic to specific enzyme systems and mitochondrial function. This phenomenon of sequestered nanoparticles within intracellular phagocytic compartments and endosomal vesicles plays a major role in decreasing nanoparticle excretion in vitro, as well as following injection of nanoparticles used in biodistribution studies in vivo.

Many investigators have found that direct contact between nanoparticles and cell surfaces, whether in vivo or in vitro can cause cell damage through lipid peroxidation of cellular membranes [3–5, 9, 11, 12, 15], the formation of reactive oxygen species (ROS) [3, 15–17], increased production of specific cytokines [3, 16, 18], and the genetic up- and down- regulation of specific proteins involved in inflammatory processes, and abnormalities in cell-adhesion [10]. Yet other investigators have found that specifically engineered nanoparticles can be designed that cause no apparent or significant permanent damage to biological cells or tissues studied following exposure [4, 12, 19, 20]. The key in developing an engineered nanoparticle for maximum biocompatibility (for the desired application) is '**testing**'. The in vitro approach for investigating cell interactions with specific nanoparticles offers a quick and less expensive way to study cellular responses to specific nanoparticle doses, exposure times, intracellular nanoparticle fate with specific cell types, and the ability to limit extraneous environmental variables as needed. Many investigators use this form of testing because it permits biochemical, ultrastructural, and molecular genetic/ proteomic analysis of specific types of cells during and following nanoparticle exposure, which is simply not possible with the complexity and variables inherent with living organisms. The obvious drawback to this approach is that it only provides data for one or a few types of cells, when in reality, nanoparticles in vivo will interact with many tissues and types of cells. Therefore, in vitro testing is more accurate and appropriately used to study specific variables and cellular responses in detail.

3 Carbon-Based Nanomaterials

Carbon nanoparticles and carbon nanomaterials are currently being utilized commercially for numerous applications in electronics, reinforced structural materials, sports and camping equipment, microfabricating conjugated polymer activators, biosensors, ultrasensitive chemical and physical sensors, solar cell and hydrogen storage technologies, cosmetics, paint, tires and more recently in biomedical imaging and therapeutic treatment modalities [4, 17, 21]. With the millions of tons of carbon nanotubes, nanohorns, nano-rings/loops, graphene sheets, fullerenes and carbon black that collectively have been, and are currently being, manufactured in academia and industry, there is a great need to understand how these uniquely adaptable nanoparticles that have such unique characteristics for utilization [22] in so many applications, may impact those who manufacture and study them, as well as impact the consumer, and the environment; where disposal may affect the groundwater, presenting risks to human and animal health and safety.

Because of the numerous types of carbon nanoparticles, this tutorial will focus on the responses of cells and organisms to manufactured 'as prepared' carbon nanotubes and purified carbon nanotubes following acid cleaning to remove metal catalyst particles. For industrial, medical and research purposes, these are the two most widely used forms of carbon nanoparticles, with fullerenes following close behind. "As produced" carbon nanotube (CNT) material is a complex mix of carbon nanotubes, bundles of single walled nanotubes (SWCNTs), thick nanoropes, nontubular carbon (graphene sheets, carbon black) and metal catalyst particles [23,24]. Most often for research and commercial uses, these carbon nanotubes are cleaned by various methods to reduce the metal catalyst, amorphous carbon and unwanted components from the "as prepared" material (Fig. 1).

Manufacturers of carbon nanoparticles produce carbon nanotubes with different combinations of metal catalyst particles (Ni, Y, Co, Al, Fe, etc.). It is also possible to get purified carbon nanotubes devoid of metal catalyst, other forms of carbon as well as the residual graphene sheets, carbon black, and amorphous carbon found in the 'as prepared' material. Metal-free nanotubes can be cleanly produced using an

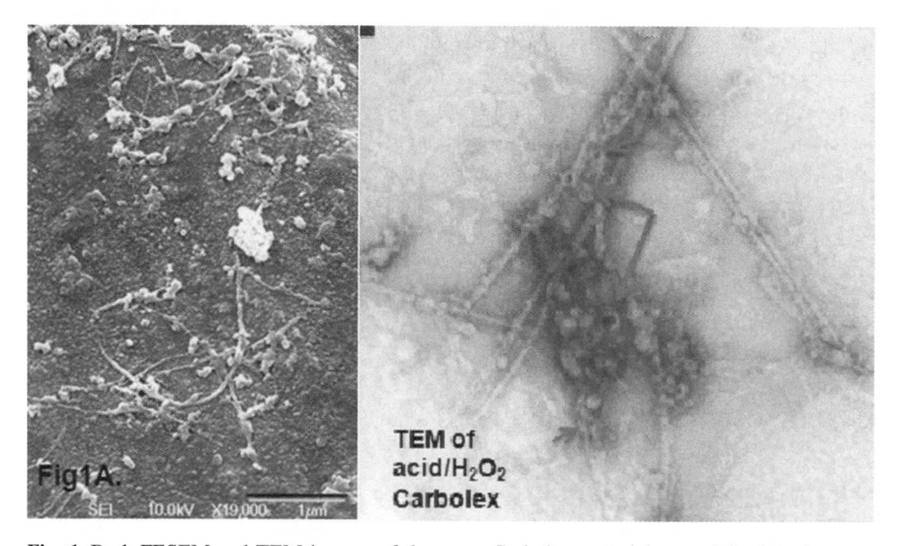

Fig. 1 Both FESEM and TEM images of the same Carbolex material treated for 97 minutes to remove debris, Ni and Y metal catalyst particles (*arrows* show remaining small amounts of metal catalyst in unstained TEM image). Some SWCNTs (thinner filaments in both images) can be seen associated with thicker nanoropes and some remaining globular material

arc discharge synthesis method [22]. Or alternatively, 'as prepared' nanotube preparations can be cleaned using chemical protocols which may also involve oxidation with O_2, H_2 or air [25].

To fully assess the scope of biological interactions, and assess the risks that these materials may pose to manufacturers, researchers and those handling or working with nanotubes in the biomedical fields (for diagnostic or treatment purposes), as well as the implications for environmental contamination and human health and safety, it is necessary to study the cell-nanoparticle interactions with the 'as prepared' and highly purified carbon nanotube preparations. Multiwalled carbon nanotubes have been found in culture to invade cells and become associated with cellular organelles and nuclei, suggesting a high probability of cytotoxicity [3,4,13,17]. Here, we will focus in this tutorial on the SWCNT, rather than MWCNTs, fullerenes and other carbon nanoparticles. To date, there have been many studies that have demonstrated that SWCNTs under specific conditions, both in vitro and in vivo, can be biocompatible, and can also cause cytotoxicity.

3.1 Experimental Methods

I developed a human epithelial cell model for studying 'cell-nanoparticle' interactions *in vitro*, in which human lung and human colon cell monolayers (Caco-2 and NCI-H292 cells) were grown to confluency and incubated with Carbolex carbon nanotubes (Carbolex Inc., Lexington, KY). Carbon nanoparticles were suspended in sterile phosphate buffered saline (PBS) to produce pH 6.8 solutions of 10 μM and 100 μM concentration (based on the carbon content). Cell monolayers were grown on 12 mm coverslips to 95–98% confluency (incubation at 36°C) in individual shell vials with 1.5ml of low serum culture medium, and to each vial was added 2 μl of a carbon nanoparticle solution (eaither at 10 or 100 μ M contraction). The vials were swirled during nanoparticle exposure to assure mixing and distribution of the nanoparticles during the incubation period of 2–4 hr (Fig. 2). Control cell monolayers were incubated in the same low serum culture medium for the same time periods without any pretreatment, or they were given 2 μl of sterile PBS prior to incubation.

Viability measurements were made using 0.4% erythrosin B vital stain (1:5 dilution with PBS) to identify necrotic cells following incubation. This stain was used rather than Trypan blue or other viability stains because it reacted rapidly (within a few seconds), was very selective for only necrotic cells, was easily photographed for counting statistics, and had been found to produce no cell toxicity (allowing for continued growth of cells following use) [26]. Cell monolayers were washed two times with warm PBS, and the cells stained with erythrosin B for several seconds, followed by a PBS rinse to remove residual stain and photographed for statistical counting. Following incubation, the medium was removed from each shell vial and the cell monolayers were washed two times with warm PBS. The wash solution and medium were then centrifuged for 20 minutes at 3400 rpm. When centrifugation

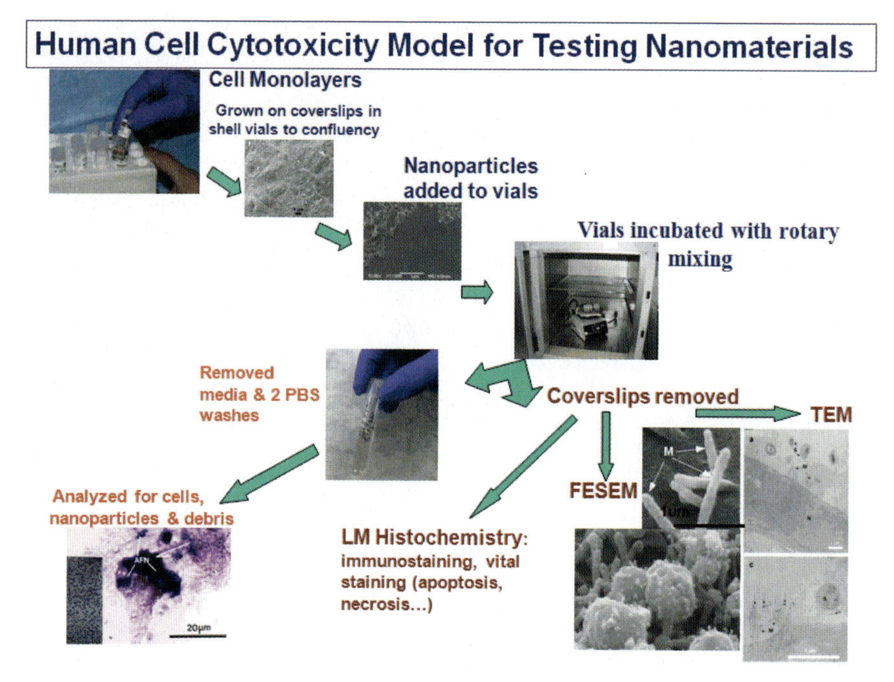

Fig. 2 Flow diagram of cytotoxicity model. Human cells grown on coverslips in individual dram vials were given 1 or 2 μl of nanoparticle suspension. The cells were then incubated at 36°C with continual mixing. Following incubation, the culture medium from each vial was analyzed, and cell monolayers prepared for viability testing, TEM and SEM imaging

produced a visible pellet, this material was prepared for light microscopy (slides) to identify whether the pellet contained contaminants, debris, nanoparticle aggregates or detached cells from the monolayer. Two slides were made from the pellet material, and one was stained with methylene blue/azure II to identify cellular debris and detached cells, and the second slide stained with Gram stain to screen for microbial contaminants. The supernatants of the media and buffer washes were analyzed by UV-vis and TEM to screen for nanoparticles.

TEM: Cell monolayers prepared for TEM were fixed in 3% glutaraldehyde in 0.1 M cacodylate buffer with added $CaCl_2$and 10% sucrose (pH 7.2). Cell monolayers were rinsed two times in 0.1 M cacodylate buffer with 10% sucrose and post fixed in 1–2% osmium tetroxide in 0.1 M cacodylate buffer. The coverslips with attached cell monolayers were then dehydrated incrementally in acetone and infiltrated with epoxy resin and polymerized at 56°C for 2 days. The tissue was thin-sectioned for TEM imaging using serial sectioned unstained, and uranyl acetate/lead citrate stained sections. We imaged the stained sections, and compared these images to the companion unstained sections to clearly differentiate between carbon nanoparticles, electron dense stained intracellular membranes and organelles, and possible staining artifacts.

FESEM: For field emission scanning electron microscopy (FESEM) using a JSM-6500F microscope equipped with an x-ray detector, cell monolayers were fixed in glutaraldehyde and post-fixed with osmium tetroxide as previously described

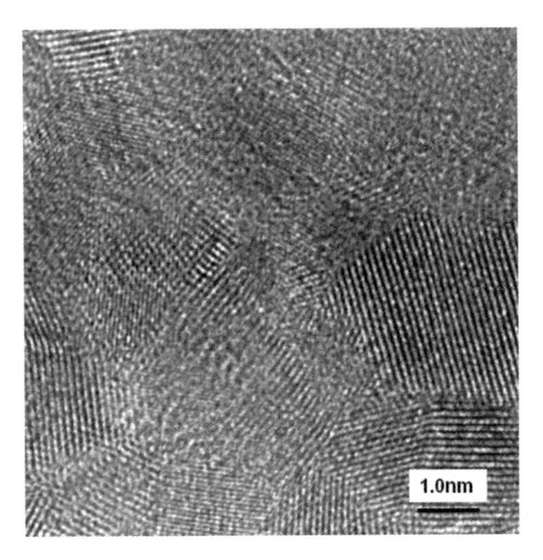

Fig. 3 TEM image of Pt coating used to coat our FESEM samples. These sputter coated 3–4 nm Pt coatings were uniform, continuous and polycrystalline, providing us with a virtually structureless coating for imaging the human cells and nanoparticles. Pt coating on SiN_3 film imaged at 300 kV

to insure optimum preparation of the membranes and cellular fine structure. Cell monolayers were dehydrated in acetone and dried by the critical point method with liquid CO_2. Monolayers were attached to specimen holders with silver paint and the samples were given a 3–4 nm thick platinum (Pt) coating with an Emitech K575XD Turbo sputter coater (Emitech Products, Inc., Houston, TX) with a peltier cooled stage to protect the specimen from heat damage and decoration artifacts. We found that this type of plasma coating produced a Pt coating of the specimen surface (and sides) that was a continuous uniform crystalline coating (Fig. 3).

To examine nanoparticles on human cell surfaces, we needed a virtually structureless coating that provided a continuous surface for heat and electron conduction, and sufficient secondary electron production for imaging at very high magnifications. We analyzed the continuity of the Pt coating produced by our turbo-pumped plasma coater by placing a silicon nitride film next to our FESEM samples within the sputter coater, and delivered the same Pt coating to the cell preparations and the SiN_3 film. We then imaged the Pt coated SiN_3 film using a JEOL 300 TEM at 300 kV and found that the Pt coating was polycrystalline and continuous. The electron diffraction patterns obtained from these films were face centered cubic. Therefore, the small structures seen in the FESEM images in both this section and the section on gold nanoparticles are not due to coating artifacts, but represent either cellular material (proteins, mucus or subcellular fine structure) or nanoparticles. Since all samples were studied by TEM and FESEM, we were also able to compare our FESEM images with the sectioned material of identically prepared cells for verification of structures.

3.2 Viability and Microscopy Results

Viability Studies: The 'as prepared' Carbolex (Carbolex Inc., Lexington, KY) that was used in these experiments had predominately SWCNTs with some nontubular graphene, Carbon black, amorphous carbon, and metal catalyst (Ni and Y) particles. This Carbolex material was air oxidized to remove much of the metal catalyst (by x-ray microanalysis, this procedure removed all Y and most but not all of the Ni metal) and all but the amorphous carbon and nanotubes. Into each dram vial containing a 12 mm coverslip (with either Caco-2 cells or NCI-H292 cells grown to 95–98% confluency), 2 μl of either 10 μM or 100 μM of carbon nanoparticle suspension in PBS was added. Compared to the control values, lung cells incubated with the 10 μM 'as prepared' Carbolex and air oxidized Carbolex showed no increase in cellular necrosis at 2 hour incubation (Fig. 4). Even the 100 μM 'as prepared' Carbolex showed no increase in lung cell necrosis at 2 hour incubation compared to control values, and only the 100 μM air oxidized Carbolex showed a small increase in cell necrosis. At 3 hour incubation, however, both 100 μM carbon nanoparticle solutions caused significant cell death (Fig. 4), but the lower 10 μM dose produced no significant cytotoxicity.

Colon cells incubated for 3 hours with either the 10 μM or the 100 μM 'as prepared' Carbolex solution showed the same high number of dead cells compared to Control values. Although at 2 hour incubation the 10 μM concentration of air oxidized Carbolex produced no increase in cell death, the 'as prepared' Carbolex caused cell necrosis similar to the 100 μM concentration. The air oxidized Carbolex preparation (which contained significantly less metal catalyst), also revealed a significant increase in cytotoxicity for both the 10 and 100 μM nanoparticle concentration at 3 hour incubation. Colon cells seemed to be more sensitive to these carbon nanoparticle preparations, even at the lower concentrations. Although the 10 μM air 'oxidized Carbolex was not cytotoxic at 2 hours incubation, at 3 hr incubation the air oxidized 10 μM dose produced the same high cytotoxicity (cell death) seen following colon cell exposure to the 'as prepared' Carbolex and air oxidized Carbolex aat the 100 μM doses.'

The lung cells were much more tolerant than the colon cells to both preparations of Carbolex at the 2 hour exposure time, and at the lower dose at 3 hours. However, significant cell necrosis occurred in both types of cells following 3-hour exposure to the 100 μM air oxidized Carbolex and 'as prepared' Carbolex, indicating that the toxicity could not be linked to the presence or absence of metal catalyst. **The different responses to nanoparticle exposure of these two different types of epithelial cells reveals how the different genetic programming of cell function, receptors, surface coatings and cell responsiveness can alter each type of the cell's interaction with specific nanoparticles. Therefore, it is not possible to assume that a nanoparticle that is very biocompatible with lung cells might be equally biocompatible with other types of cells. And conversely, it is a mistake to predict that an acutely cytotoxic response using spleen or thymus cells would result in the same highly toxic response in keratinocytes cells.** At this early stage in studying nanoparticle biocompatibility, experimentation with many cell types *in vitro*, and subsequent further testing in vivo are necessary to understand how specific nanoparticles will interact with any given type of cell or tissue. This biological component to nanoparticle biocompatibility is often not considered adequately, leading to assumptions that may not be true in all cases.

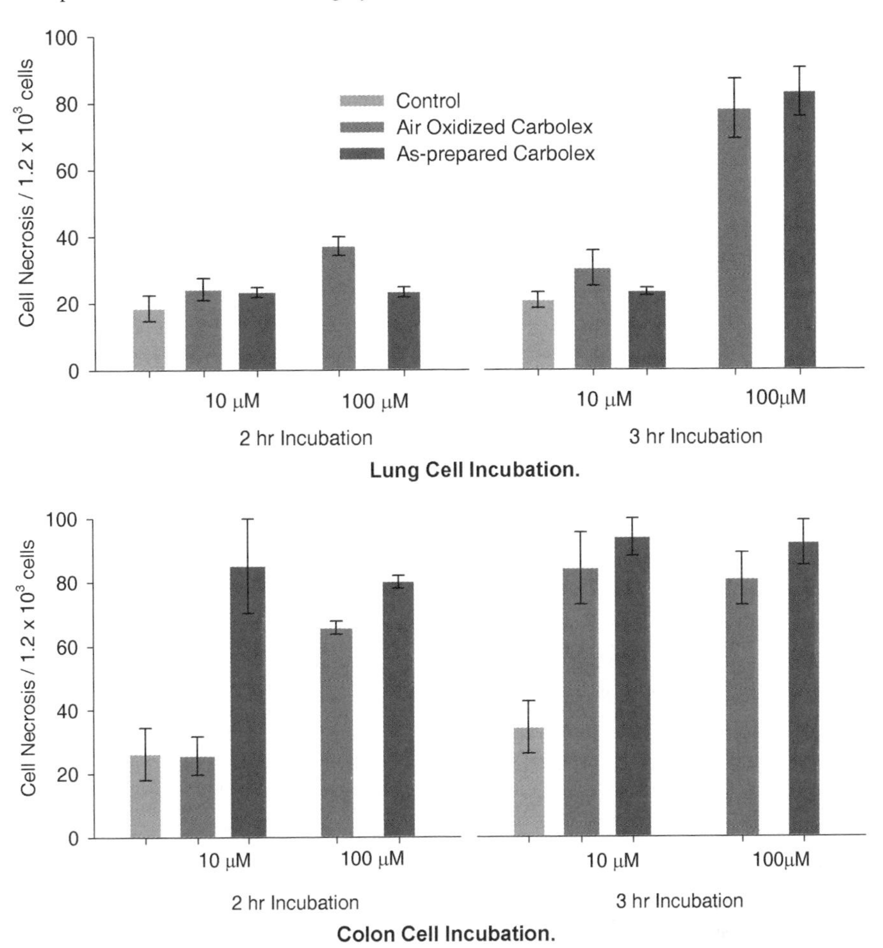

Fig. 4 NCI-H292 Lung and Caco-2 colon cells were incubated for 2 h or 3 h with 2 μl of either10 μM or 100 μM of 'as prepared' Carbolex, or Air Oxidized Carbolex solutions (pH 6.8). Following exposure, the cell monolayers were rinsed with warm PBS and stained with erythrosin B vital stain, photographed and necrotic cells counted in several fields from each monolayer. N = 16 to 19. Standard deviation ±

Microscopy: By FESEM, the cells incubated with 10 μM Carbolex showed some loss of adherence of cells and some increased cytoskeletal activity indicative of stress (Fig. 5). At 3 hour incubation, the damage was only slightly worse in the 100 μM exposed cell monolayers than the 10 μm exposed monolayers, which matched the viability results for 3-hour colon cells. The monolayers exposed to the higher concentration of 'as prepared' Carbolex showed disrupted cells with damaged nuclei and segments of the cells missing, and many detached cells with large open areas of the coverslip bare(Fig. 5A). Although the 10 μm exposed cell monolayers retained more of their individual cell morphology with many cells still

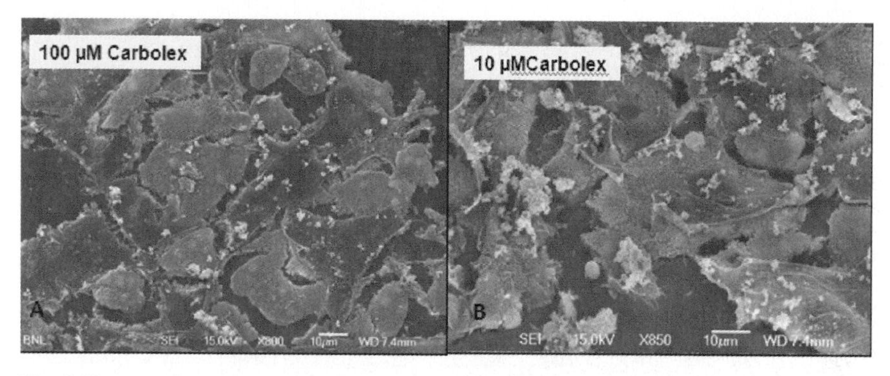

Fig. 5 Human colon cells grown on glass coverslips and incubated for 3 hours with 10 or 100 μM solutions of 'as prepared' Carbolex in PBS (pH 6.8) showed missing cells in the 10 μM exposed cell monolayers as well as some cell damage. The 100 μM Carbolex exposed monolayers showed cell fragmentation, loss of almost all cell-to-cell adhesion and the loss of cell-to-substrate adhesion

intact, many of the cells were missing (loss of cell adhesion) and the cells no longer displayed cell-to-cell contact, suggesting decreased E-cadherin and adhesin production (Fig. 5B). This loss of cell adhesion to the coverslip and to adjacent cells in the monolayer has been described as indicative of nanoparticle cytotoxicity [3, 5, 10].

The Caco-2 colon cells produced a mucous-like material on the apical surface when incubated with the Carbolex nanoparticles. Carbon nanotubes that initially attached to the apical mucus within the 2-hour incubation period did not initially seem to cause any problems for the cells (Fig. 7). However, once the carbon nanopar-

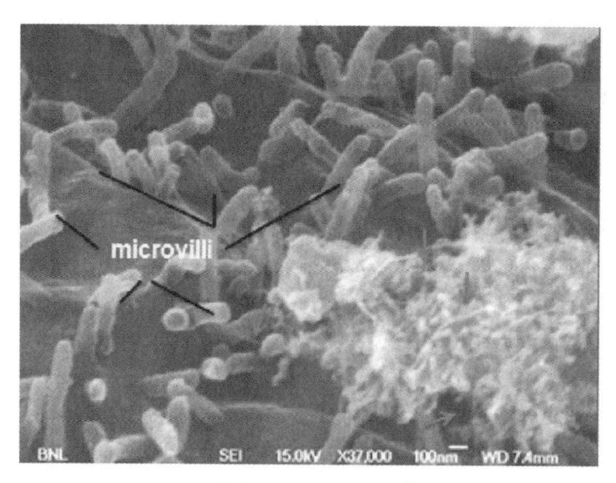

Fig. 6 Caco-2 colon cell surface with numerous microvilli and a large aggregation of carbon nanotubes, nanoropes (*arrows*) and catalyst entrapped in surface mucus. The initial response of these human colon cells when incubated with the Carbolex preparations was to produce clusters of mucoid material on the apical cell surface. However, at this time period, the microvilli and the apical plasma membrane appear normal with smooth surface membranes. Nanotubes and nanoropes trapped in mucus (*arrows*)

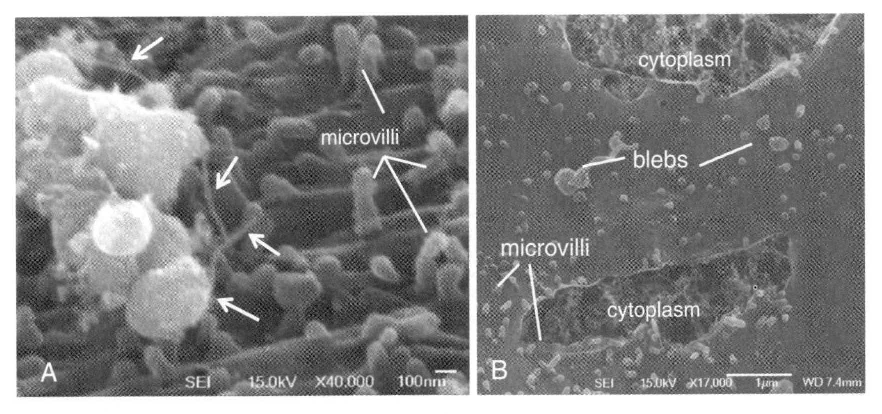

Fig. 7 Colon cells incubated with 100 μM Carbolex for 2 h (**A**) and 5.5 h (**B**). SWCNTs and nanoropes (*A, arrows*) entangled in surface mucoid material were trapped against the apical plasma membrane and microvilli at 2 h (A). Later at 5.5 h exposure (B), those cells with attached carbon nanoparticles exhibited membrane blebbing and openings in the surface apical membrane revealing the underlying cell cytoplasm

ticles made contact directly with the plasmalemmal surface (Fig. 6), the surface microvilli and plasma membranes showed eventual signs of damage. Since these experiments were all done with the nanoparticles being continually swirled in the culture media over the cell monolayers (to increase nanoparticle distribution over the entire monolayer surface), most areas of the monolayers surveyed using microscopy showed an even distribution of SWCNTs over the monolayer surface. We feel that gently swirling the media containing the nanoparticles more closely simulates the conditions that naturally occur in the gut and lung, where peristalsis and respiratory movements produce a constant mixing and dispersion of materials in vivo.

Contact between the Carbolex nanoropes and carbon nanotubes (seen in Fig. 7, arrows) directly with the cell surface caused severe damage to the plasma membrane producing holes and tears suggestive of lipid peroxidation damage, and ROS damage, reported in the literature. Where the Carbolex material touched microvillar membranes during the 3–4 hour exposure time, the microvilli revealed swelling of the distal portion of their plasma membranes (Figs. 8, 9). Exposure (5.5 h) to 100 μM 'as prepared' Carbolex caused the plasma membrane with the attached microvilli to peel away from the underlying connective fibers, thus separating the plasma membrane from the underlying cytoplasm (Figs. 7 and 8). Cellular damage was so severe at 5.5 hour exposure to 100 μM Carbolex (and viability measurements difficult to calculate with so many detached and fragmented cells), that we did not use this extended time period further for these experiments.

We also found that areas with carbon nanotubes or aggregates of carbon nanotubes touching the apical plasma membrane of the colon cells had small breaks in the surface membrane where electron dense carbon and SWCNT particulates entered the cells, passed directly through the cytoplasm and often were lodged in the nucleus of the cell (Fig. 10). At this time, it is not clear if these carbon nanoparticles

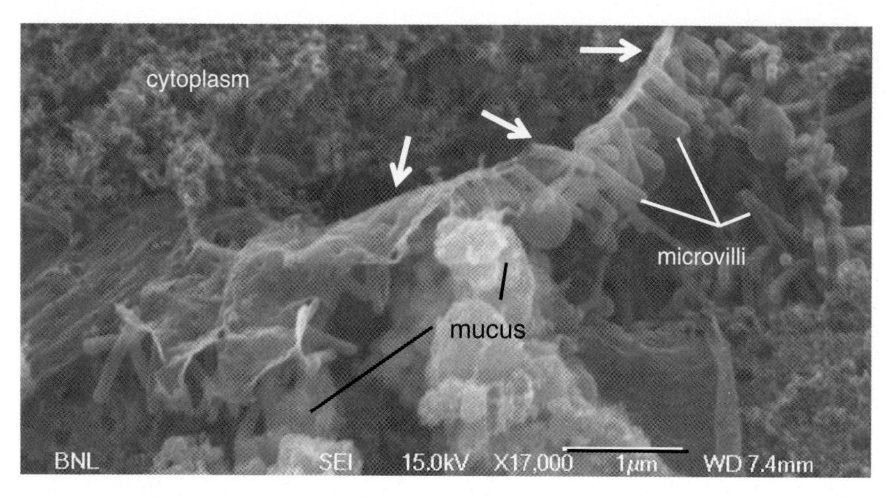

Fig. 8 Colon cells incubated for 5.5 h with 100 μM Carbolex in PBS showed a peeling away (*arrows*) of the plasma membrane (with attached microvilli), from the underlying cytoplasm due to prolonged contact between Carbolex nanoparticles and the plasma membrane. This could represent membrane damage from ROS and lipid peroxidation [10–12]

passed into the cells because the cells had contact damage to the apical plasma membrane producing small open holes, and were in the process of dying, or, if these cells could have repaired the membrane, recovered and lived with carbon nanoparticles within their nuclei.

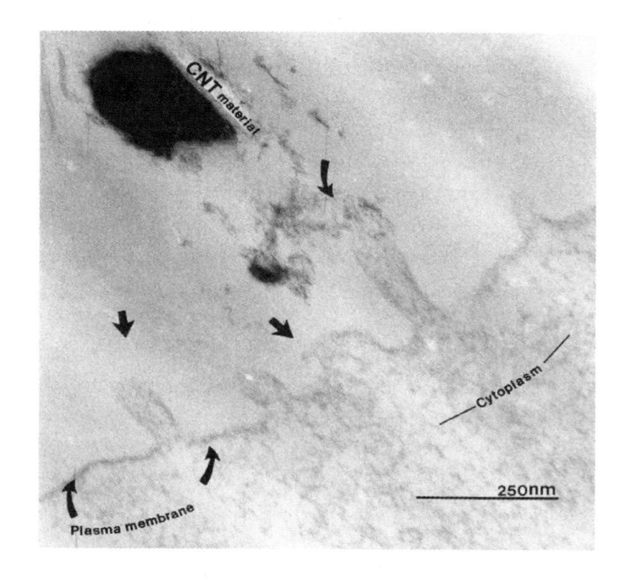

Fig. 9 TEM imaging of colon cells following 100 μM Carbolex incubation revealed that CNT-microvillar contact with aggregates, as well as individual carbon nanotubes, produced the dissolution of the microvillar membranes (*top arrows*)

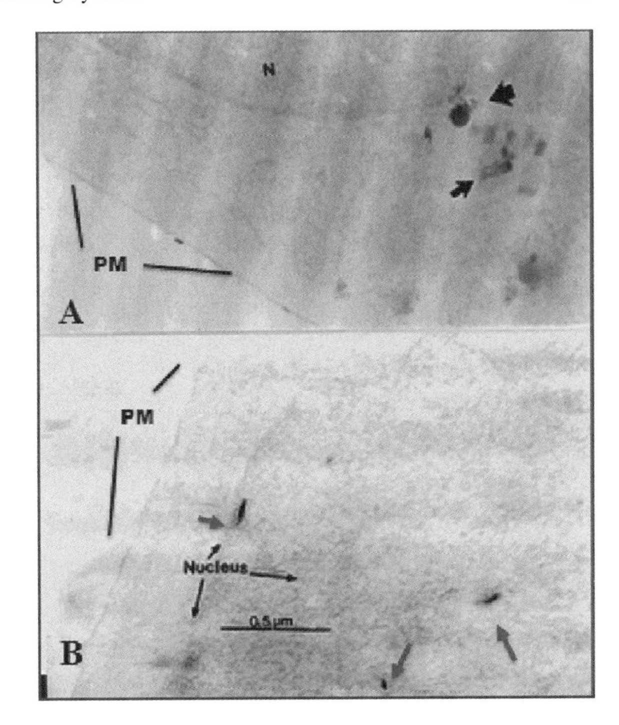

Fig. 10 A. TEM of colon cell exposed to 100 μM Carbolex showing intact nucleus (N) with carbon nanomaterial (*arrows*) free within the cell cytoplasm, adjacent to the nuclear membrane. Plasma membrane (*PM*). Figure 10B. These cells showed electron dense surface performations of the apical plasma membranes (PM) and also the occassional electron dense nanotube bundles (*largest arrows*) within the nucleus of the cells. At very high magnification, these nuclear inclusions revealed delicate bundles of SWCNTs

3.3 Carbon Nanoparticles Residing Within Living Biological Cells

Shvedova et al. [27] demonstrated that Fe enriched SWCNTs had a significantly more cytotoxic effect on cultured keratinocytes and bronchoalveolar epithelial cells than SWCNTs treated with iron chelators. They found that the presence of nanometer diameter Fe particles in commercial 'as prepared' HiPco carbon nanotube preparations produced Fenton-type reactions and the release of reactive oxygen species (ROS) when incubated with primary keratinocytes in vitro. When similar 'as prepared' HiPco SWCNTs (comprised of 99.7% carbon and 2.3% iron) and purified non-functionalized HiPco SWCNTs were incubated with macrophages, the phagocytic cells did not effectively engulf and incorporate the purified SWCNTs [17]. Using RAW 264.7 macrophages, the same HiPco 'as prepared' (containing Fe), and purified (lacking metal catalyst) SWCNTs did not stimulate intracellular production of superoxide radicals or nitric oxide [28]. These macrophages had poor recognition of the non-functionalized SWCNTs, and consequently did not actively target them for engulfment or phagocytosis, which was dissimilar to the response of the keratinocytes exposed to the 'as prepared' SWCNTs. Not only do living cells seem to suffer cytotoxicity when exposed to carbon nanotubes with metal catalyst, but different cells were shown to respond very differently to the same carbon nanoparticles. The macrophages did not actively engulf or respond cytotoxically to the carbon nanotubes, but the keratinocytes showed stress and release of ROS.

More recently Shvedova et al. [17] described two types of responses in mouse lung following exposure to SWCNTs dependent on nanoparticle aggregation or dispersion. When HiPco SWCNTs (containing iron) were dispersed during administration in mouse lung, the SWCNTs caused interstitial fibrosis in the lung. This occurred without the typical inflammatory chronic infiltration of inflammatory cells and focal granulomas seen when SWCNTs were allowed to enter the lung as aggregates using an in vivo tracheal instillation protocol [29, 30]. With the dispersed SWCNTs, the aspiration of the dispersed nanotubes did cause a rapid increase in BAL levels of inflammatory cells, inflammatory cytokines and protein, but the response resolved quickly and did not persist for the 2-month post-exposure period [17]. This was consistent with the findings of Kagan et al. [28] who demonstrated low reactivity of these SWCNTs with lung macrophages (sparse macrophage uptake of SWCNTs), and the low cytokine and superoxide generation. In vivo, not only did SWCNT exposure time and dose significantly affect lung damage [29, 30], but **the amount of metal catalyst present** from the SWCNTs and the **dispersion or aggregation** of those SWCNTs all contributed to cytotoxic responses. Shvedova et al.'s work also [17] revealed that SWCNTs could penetrate interstitial tissue and that the possibility of CNT **translocation** in vivo into the systemic circulation with ultrafine carbon particles reported in 2002 by Oberdorster et al. [31], was a very real consideration in mammalian lung as well.

Wick et al. [32] found that '**agglomeration**' affected CNT cytotoxicity with human SSTO-211 H cells and produced pronounced cytotoxic effects to the cells following exposure to rope-like CNT agglomerates, when compared to cell exposure with dispersed CNT preparations. However, Pulskamp et al. [33] found that rat NR8383 macrophages and lung A549 cells incubated with commercial 'as prepared' SWCNTs, carbon black and MWCNTs, as well as acid cleaned SWCNTs, showed no acute cell cytotoxicity (viability) with all CNT samples, nor produced inflammatory mediators (no increase in TNF alpha, IL-8 or nitric oxide). They found a dose/time dependent increase of intracellular reactive oxygen species and a decrease of mitochondrial membrane potential with commercial CNTs in both cell types. The purified CNTs had no deleterious effects on the cells, and their overall conclusion was that the metal catalyst traces in the commercial CNTs were responsible for adverse biological effects. In contrast, Tian et al. [34] reported that the surface area of refined carbon nanomaterials predicted the potential for cellular toxicity, and that refined SWCNTs (with metal catalyst removed) induced the strongest cellular apoptosis/necrosis responses in human fibroblast cells *in vitro,* and were more toxic than their unrefined comparable SWCNTs.

3.4 Carbon Nanotube Summary

In summary, SWCNT interactions with cells are not simple, nor are they easily predicted. There are many factors that contribute to cytotoxic responses, and the biocompatibility of a SWCNT is very different depending on the type of cell exposed, the length of exposure, dose, and nanoparticle used (was the SWCNT in the 'as prepared' condition, functionalized, oxidized, or cleaned & cut producing more surface reactive sites). Carbon nanotube incorporation into some cells seems to be a size-dependent phenomenon as demonstrated by Becker et al. [35] using human lung primary IMR-90 cells in vitro. The latter primary (non-cancerous) fibroblast cells were observed to incorporate DNA wrapped SWCNTs only when they were below 189 $\pm$ 17 nm in length. Identical DNA-wrapped SWCNTs measuring 335 $\pm$ 27 and 253 $\pm$ 26 nm did not enter the cells after 16-hour incubation, but remained in the media and did not produce cytotoxic effects. The shorter DNA-wrapped nanotubes that did enter the IMR-90 human lung fibroblasts by TEM were found within the cytoplasm of the cells and produced cytotoxic effects compared to the longer nanotubes. Chirality of the nanotubes was also found to play no role in cellular uptake. When Becker et al. tested different cell lines (human alveolar basal epithelial cells, A549; clonal murine calvarial cells, MC3T3-E1; and embryonic rat thoracic aorta cells, A10), the cells were similarly tested and showed the same cytotoxicity and uptake of the smaller ($>$189 $\pm$17 nm) DNA-wrapped SWCNTs, as that seen with the lung fibroblast cells. The size of the DNA-wrapped nanotubes played a major role in uptake and apparently cytotoxicity. The DNA-wrapping of the nanotubes may have facilitated cellular uptake of the smaller SWCNTs as well. Our own work with DNA-functionalized gold nanoparticles revealed that nanoparticles were rapidly bound to the plasma membrane and incorporated into both human colon and lung cells following 2-hour exposure. However, there was no visible evidence of endocytic vesicle formation, or some form of phagocytosis. These DNA functionalized gold nanoparticles were found in the apical cytoplasm, or localized within endoplasmic (ER) reticulum-like tubular channels at the apical surface [5]. They were later seen within the Golgi and finally within vacuoles in the central and basal portions of the cells. This may represent some form of DNA-facilitated transport that bypassed the usual endocytic pathway.

We found that SWCNTs, as well as other components of 'as prepared' Carbolex material, attached to both lung (NCI-H292) and colon (Caco-2) cells via cell elaborated material on the apical epithelial cell surface. Once attached to the apical cell surface, the SWCNTs only produced damage when they came in contact with the plasma membrane. When the SWCNTs or nanoropes touched the plasma membrane, or the microvillar membranes, they caused severe damage to the membrane structure producing areas of dissolution suggestive of the lipid peroxidation described by Sayes et al. [12] and others in regard to fullerene damage to cell membranes. Small carbon nanotubes and nanotube bundles also entered the cells through these breeches of the plasma membrane. By TEM, the carbon nanoparticles entered the cell cytoplasm directly and were not surrounded by a vesicular membrane, indicating a similar type of incorporation to that described for uptake DNA-wrapped SWCNTs reported by Becker et al. [35]. We also found very small bundles of carbon nanotubes and other carbon nanoparticles at the nuclear membrane and within the nucleus proper, with concomitant increased cell mortality in these samples, similar to Becker et al.'s findings.

Porter et al. [36] studied the uptake of HiPco SWCNTs in human monocyte-derived macrophages treated for 2–4 days with 0–10 $\mu g\,ml^{-1}$ SWCNTs dispersed in tetrahydrofuran. They found that the majority of SWCNTs were located in phagosomes and lysosomes suggesting phagocytosis. From 4 days' incubation, SWCNTs were seen in the cytoplasm which they felt indicated "passive uptake through the lipid bilayer" and they observed both apoptotic and necrotic cell death, with cell mortality corresponding to areas of high SWCNT density. Nanotubes were seen within cell nuclei after 3 days, and SWCNTs touching the lysosomal membrane were reported to cause membrane disruption. It should be noted here that monocyte-derived macrophages are professional phagocytes, and that they are specialized cells that engulf cell debris, foreign and microbial material, in organs and the blood stream. Materials can remain within phagocytic vacuoles intracellularly for quite some time without rupture or transport out of the vacuole. Therefore, the loss of cellular compartmental integrity within these cells from day 4 on, by the passage of nanoparticles through intact intracellular membranes via damage to those membranes, suggests a very toxic response. It may also be that the cells were already dying at this time (since cell death was reported from day 4 on), and consequently the membranes were no longer functioning, allowing breaks to appear at regions with accumulated SWCNTs. Because the authors also used HIPco nanotubes which are known to be more toxic to cells [17], it is not clear if the carbon nanotubes had adherent Fe particles associated with them which could have increased cytotoxic effects.

4 Radiolabeled Quantum Dots and Shell Cross-Linked Nanoparticles

4.1 Quantum Dots (QDs)

Studies using various types of quantum dots in vivo (amphibians, rodents) have shown that QDs can be well tolerated and produce clear imaging for biodistribution studies and targeting of specific organs [19,20,35–39]. Ballou et al. [37] found that circulating half lives of the CdSe QDs with a zinc sulfide shell (CdSe/ZnS) varied with the surface coating, with less than 12 minutes for amphilphilic poly (acrylic acid), short-chain (750 Da) methoxy-PEG or long-chain (3400Da) carboxy-PEG coated quantum dots; and $\sim$70 minutes for long chain (5000Da) methoxy-PEG quantum dots. The surface coating on the QD also determined the in vivo localization and increased biocompatibility of the nanoparticles. Cadmium/Se quantum dots coated with PEG-750 injected into mice were imaged at 15 minutes, 1, 3, 7 and 28 days. Even at 1 month post-injection, these QDs were found at necropsy in liver, lymph nodes and bone marrow. At 4 months post-injection, QDs were still evident in lymph nodes, spleen and liver, but they had produced no apparent pathology in the day-to-day life of the test animals [37]. Similarly, Jaiswal et al. [40] reported that tissue culture cells containing QDs showed no diminished cell growth or altered cell division during weeks of incubation. In in vivo studies using mice, CdSe/ZnS polymer-coated QDs produced no deleterious effects to the treated mice even after months post-injection [9]. In *Venous* embryos, Dubertret et al. [41] were able to image QD-phospholipid micelles throughout embryo development without any apparent alteration in the animals' phenotype. Although these findings suggest that these QDs are biocompatible, the fact remains that not all of the quantum dots were excreted from the organisms or cells even after many months post-injection.

This phenomenon of **bioaccumulation** can produce significant problems to cells if the biocompatible coating and/or the ZnS shell of the QD breaks down from cellular enzyme or acid degradation (lysosomal attack), and the cadmium and selenium are released into the cells and tissues. Cadmium and Selenium are both toxic to cells in sufficient concentrations, and even low levels of Cd ions (100–400 μM) have been shown to decrease viability of hepatocytes in vitro [42]. Intracellularly, even small amounts of Cadmium will bind to sulfhydryl groups of mitochondrial proteins causing thiol group inactivation which leads to oxidative stress and mitochondrial dysfunction [42,43]. Using hepatic primary cells in vitro, Derfus et al. [42] demonstrated that QDs were cytotoxic under certain conditions (exposure to ultraviolet light and altered or oxidized surface coatings), and that release of free Cd ions correlated with cell death (Fig. 11). The stability of the QD outer coating determined whether free Cd^{2+} ions would escape from the QD core and cause hepatic cytotoxicity. Their findings verified that the surface coating played an important role in QD biocompatibility by providing a smooth, continuous outer covering which (a) completely encapsulated the CdSe core separating it from the biological environment; (b) reduced the chance of serum proteins and especially immuno-proteins attaching to the QD surface; and (c) prevented surface oxidation of the QD.

Quantum dots have been found within phagocytic cells in the spleen, lymph nodes, and the liver of rodents many months post-injection, indicating that clearance of the QDs may take some time [37,44–46]. The reticuloendothelial system (RES) has specialized cells called **tissue histiocytes** in the skin, lung, brain, liver, lymph nodes, spleen and thymus which screen the blood, lymph and cerebrospinal fluid for foreign particulates [1]. When foreign materials are found by phagocytic cells, they are sensed and engulfed and stored in phagocytic vacuoles. If these nanoparticulates were bacteria or parasites, they would be digested by the enzymes within the phagocytic vacuoles in these cells. However, these enzymes are not designed to digest polymer coatings, Cd/Se nanocrystals or any nonorganic nanoparticles in general. Therefore, the cells of the RES often retain the nanoparticles, releasing them slowly or not at all. Over time, there is concern that the low pH (pH1-2) and enzyme bombardment within phagocytic vacuoles or endosomal vesicles may hasten deterioration of the outer coating of the QD and allow Cd and/or Se release into the cell cytoplasm causing serious damage. In mammals, Cd exposure has been implicated in the intracellular production of metallothionein, and to zinc substitution, both of which are linked to pathologic conditions. Prostate, testicular, lung and pancreatic cancers are believed to be associated with Cd exposure [47,48]. "The biologic half life of Cd in humans is 15–20 yrs and it bioaccumulates, can cross the blood brain

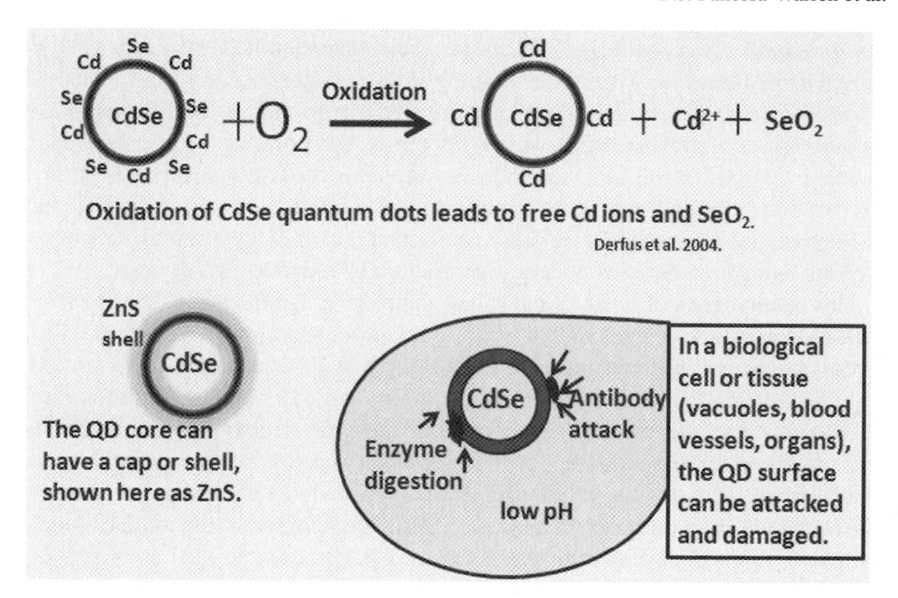

Fig. 11 Schematic diagram of ZnS capped CdSe quantum dot, and the process of surface oxidation leading to the release of Cd ions. Modified from Derfus et al. [42]

barrier, with the liver and kidney being the target organs of toxicity" [49]. See is also toxic causing death in humans at high doses. However it should also be stated here that "not all QDs have CdSe cores". Each type of QD possesses its own unique physicochemical properties which in turn determines its potential toxicity [49] or biocompatibility.

4.2 QD Methods and Results

Recent research done at Brookhaven National Laboratory by Wynne Schiffer and her colleagues [44–46] utilizing 10 nm or 2 nm diameter C^{11} radiolabeled CdSe/ZnS [50] for studying nanoparticle biodistribution using MicroPET imaging, revealed that the size of the QD made a difference in the retention of the nanoparticle within specific organs in the rodent [44]. Mice were injected (tail vein) with either 2 nm or 10 nm diameter radiolabeled [C^{11}] CdSe ZnS capped QDs and the QD distribution imaged by MicroPET for 5000 seconds. The 2 nm QDs rapidly entered the mouse brain, followed by a rapid efflux up to 1000 seconds. Thereafter, the rate of efflux slowed with some nanoparticles still remaining in the tissue at 5000 seconds. The 10 nm diameter QDs seemed to be less able to enter the brain from the bloodstream with less than 2% of the QDs entering. Those QDs that did pass through the blood brain barrier were not all able to exit the brain tissue at the end of the testing period (5000 s). Nanoparticle retention within tissues was verified by microscopy and quantitative elemental analysis of organs and tissues at necropsy (x-ray microanalysis and ICP-AES) [44, 46]. This retention of the larger nanoparticles in brain tissue may represent ingestion of the nanoparticles by brain phagocytic cells (the microglia); nanoparticle aggregation and adsorption of serum or immune proteins, triggering their attachment to endothelial cells lining the blood vessels, or macrophage; or histiocyte phagocytic engulfment of the QDs. Some recent work by Sun et al. [51] also introduced the possibility that some nanoparticles could become 'stuck' in tissues or vessels in vivo due to the lack of flexibility of the nanoparticle core during passage through pores, fenestrations, sinusoids or cell membranes. (This will be presented later in this lecture, Sect. 4.3.)

Short-term Cytotoxicity Testing: The initial results obtained by Schiffer et al. [44–46] indicated that in vivo low doses of QDs over a short period of time seemed to work well for the MicroPET imaging and the ability of the mice to tolerate the systemic administration of QDs and the procedure in general. To see if the CdSe/ZnS quantum dots produced any short-term cytotoxic effects within the 1.5-hour exposure period (comparable to the mouse exposure during MicroPET imaging); I tested both diluted and stock solution concentrations of QDs with my epithelial lung and colon cell nanoparticle cytotoxicity model. Because there has been concern in the literature about deterioration of CdSe lattice with time, and the subsequent toxic release of Cd^{2+}cations into cells and tissues, I chose to look at both freshly prepared QD preparations (labeled N for new) and previously prepared old QD stock solutions (labeled O for old). The old preparation had been prepared 8–10 months before, and was refrigerated at 4°C during the storage period. By UV-vis, the New (N) and Old (O) stock preparations were 1.47 μM (2 mg/ml). The response of the in vitro cells to the same diluted QD preparation used in the mouse *in vivo* experiments (at 0.19 μM, 0.01 mg/ml) was also tested (Fig. 12).

Cell monolayers were grown on coverslips to 94–98% confluency in 1.5 ml of low serum protein media at 36°C in shell vials. A 1-μL dose of each QD preparation was added to the individual shell vials, and all of the vials were rotary swirled during the 1.5 hour incubation to distribute the nanoparticles over the cell surfaces and reduce the chance of nanoparticle aggregation and areas of the monolayer without any QD distribution. To see if increasing the dose of the new QD preparation changed the biocompatibility of the cellular response, the QD dose was increased to 2 μl, labeled in Fig. 12 as 'Double Dose'.

The viability results revealed a difference in the responses of the lung cells compared to the colon cells exposed to the diluted (lower) QD dose and the doubled dose. The colon cells exhibited greater cell death with the diluted QD compared to the lung cells, but the lung cells showed greater cell death when incubated with the double QD dose. This disparity in cytotoxic responses between the two cell types demonstrates the importance and need to experimentally test nanoparticle compatibility with specific types of cells.

Schiffer et al. [44,45] also found that QDs remained in kidney, gall bladder, stomach and liver rodent tissues at the end of the 5000-second data collection period. When we tried to identify where the QDs might be localized within tissue samples using x-ray microanalysis, we had difficulty getting clear x-ray signal above

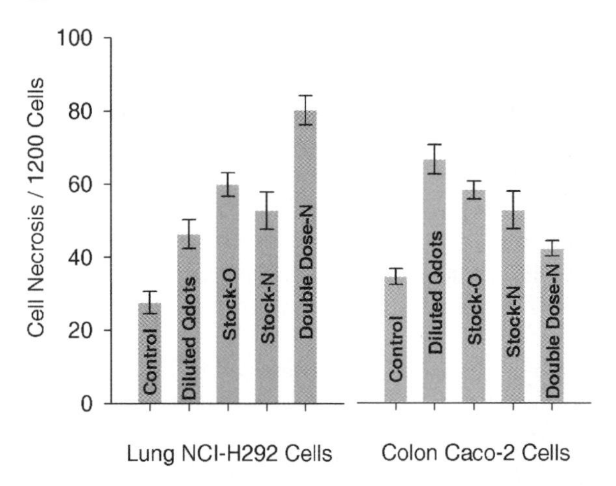

Fig. 12 Lung and Colon cells incubated 1.5 hours with high and low doses of QDs statistically showed very similar responses to the old (O) and new (N) stock solutions. Colon cells responded to the diluted QD dose with greater cell death than the lung cells. Exposure to double the dose of QDs produced the greatest cellular necrosis in the lung cells, with minimal cytotoxicity in colon cells. N = 4 to11 Standard deviation ±

background due to the very low concentrations of the QDs in any given tissue following tail vein injection, and the weaker Lα x-ray lines produced for Cd and Se (Figs. 13, 14 and 15). The calculated concentration of the diluted injected dose was 0.092 μM at 0.01 mg/ml.

X-ray spectra were taken of mouse kidney tissue slices fixed in paraformaldehyde in phosphate buffer, dehydrated in acetone, dried by the critical point method with liquid CO_2, attached to a carbon specimen mount with graphite cement, and carbon coated (4–5 nm) for x-ray microanalysis (using raster mode to collect the most QD x-ray signal above background). The calculated concentration of the injected dose of QDs given to this mouse was 0.29 μM (0.014 mg/ml), which was actually too low to clearly see Cd and Se x-ray lines in situ with a silicon detector, unless the nanoparticles were aggregated in large groups, or concentrated by biological processes in the area of electron beam excitation. Therefore, we examined areas of tissue where QDs would be in the greatest numbers (i.e. in the kidney cortex and collecting ducts of the kidney medulla, and on dense accumulations of blood cells within capillaries and arterioles. For the sake of brevity, I am only presenting kidney tissue results in these lecture notes.

Normally, blood entering the kidney passes through a filtration system in the kidney cortex, where toxins, water and various ions are filtered out of the blood into the newly formed urine. To conserve water and some electrolytes, the urine is then concentrated by the tubule cells and in the loop of Henle before it passes into the collecting ducts. The collecting ducts carry the concentrated urine into the renal calyx (a large chamber) where it passes out of the kidney via the ureters. When we looked at arterioles in the kidney cortex, we found numerous aggregates clustered on red blood cells (RBC) [44–46] (Fig. 14). These aggregations of particles were

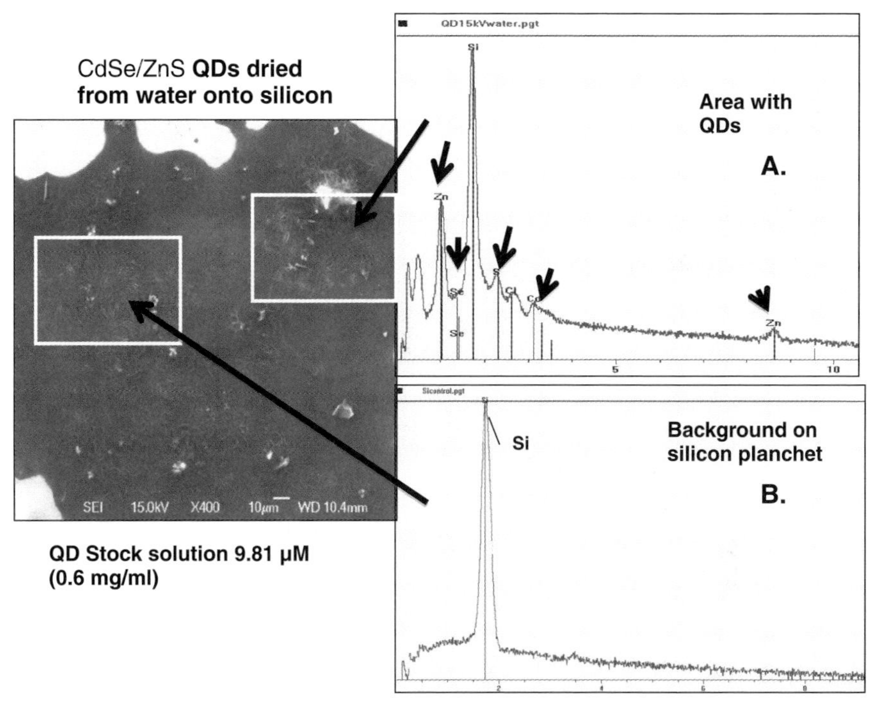

Fig. 13 Shows the SEM image of QDs (from stock solution 2 mg/ml, 1.47 μM) on a SiO_2 wafer. X-ray Spectrum A shows x-ray signal from QDs (11×10^3 counts full scale) with Zinc Lα1 and Kα, Sulfur Kα, Cd Lα1 and Lβ1 (*little arrows in the top panel on the righthand side*). High silicon x-ray signal has buried the Se L lines. In a QD free area, the Si background spectrum B, showed no Zn, S or Cd x-ray signal for the same counting period. Analyzed at 15 KV in raster mode-boxes in SEM image)

not seen on the red blood cells in control tissue or on the RBCs in other tissues nor in other areas of the kidney. X-ray spectra taken on the RBC aggregated particulates produced iron (Fe) x-ray signal at 6.4 keV (from hemoglobin), but only trace amounts of Zn, Cd and Se x-ray signal (Fig. 14 spectrum A). Spectra taken within RBC-filled cortex arterioles produced stronger QD signal. These aggregates may represent attached QDs on the endothelial wall (by complement fixation), or QDs passing into the endothelial cells. The tissues analyzed in Figs. 14 and 15 were from an animal that died shortly after the QDs passed through the kidney and into the bladder. Although anaphylaxis was suspected at the time, we did not see any platelet involvement in the blood vessels (clots), or engorged leukocytes with phagocytized QDs, which would have been characteristic of this type of acute toxicity.

Analysis of the collecting ducts and the renal calyx in the kidney medulla showed Cd and Zn (Fig. 15, spectrum A) x-ray signal *in situ*. The background spectrum on the interstitial tissue of the collecting ducts produced such high Na, P, S, Ca and K x-ray signal that it was not possible to see any of the characteristic QD x-ray peaks above background (Fig. 15, spectrum B). MicroPET imaging of the mouse kidney

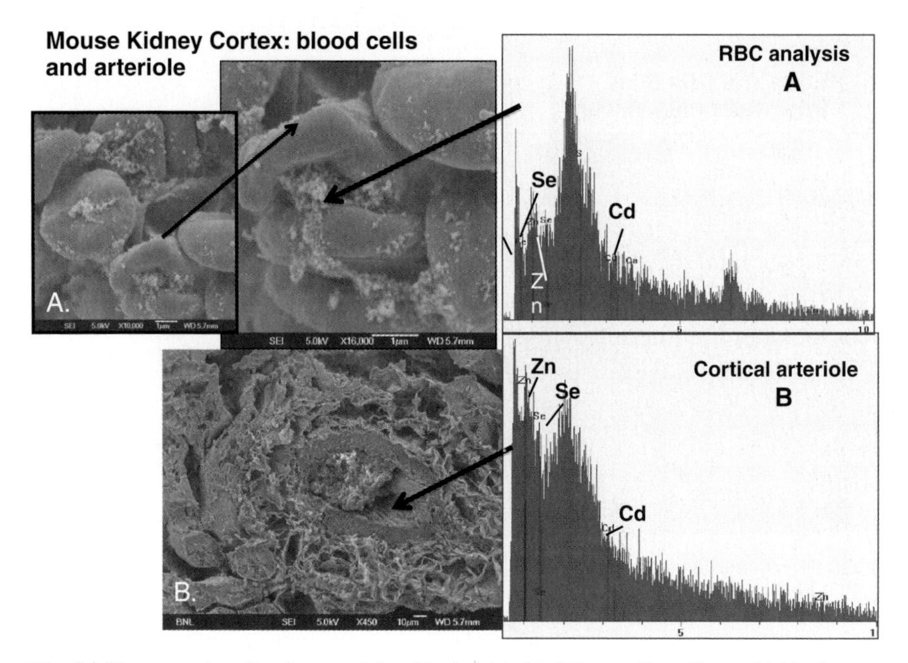

Fig. 14 X-ray spectra of carbon coated, critical point dried tissue slices (3 mm thick) of mouse kidney cortex. Dense clusters of red blood cells (RBC) (Fig. 14 A) were found within cortical kidney arterioles (14B). On the RBC surfaces were ribbons of attached aggregates that produced Zn, Se and Cd x-ray signal above background (spectrum A). Larger clusters of these aggregates were found attached to the arteriole wall endothelium, and these areas produced x-ray signal suggestive of QDs (Spectrum B).

showed that the 10 nm diameter QDs rapidly passed into the kidney medulla, exiting within 5000 seconds [44, 45]. However, the smaller 2 nm diameter QDs exited the renal medulla more slowly, with particles still retained in the tissue at the end of the testing period. In the renal cortex, the smaller 2 nm diameter QDs passed rapidly into and out of the kidney cortex (suggesting excellent clearance through the glomeruli), whereas the 10 nm diameter QDs passed out of the cortex more slowly with some retention in the tissue at 5000 seconds. Further work is in progress to identify how and where these QDs may have lodged within tissues and whether this retention of QDs is a temporary, or long term, phenomenon which may pose risks to affected tissues.

Research Summary

Similar to other QD studies in vivo, [44, 45] these radiolabeled QDs (injected dose 150 µCi in 0.4 mL solution) produced excellent MicroPET imaging for biodistribution studies and did not reveal any serious toxicity problems with the mice. Many mice were maintained after imaging for several months without signs of toxicity, yet

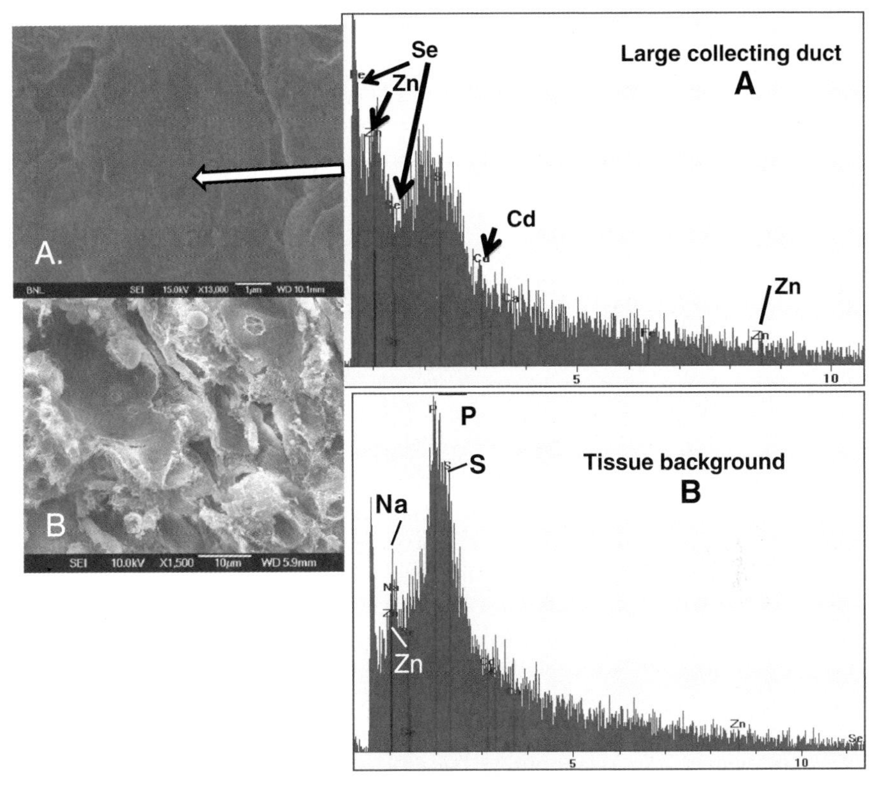

Fig. 15 X-ray spectrum (A) taken at the entrance of the collecting ducts into the renal calyx (image A) showed evidence of QDs (spectrum shows Cd, Zn and Se x-ray signal), compared to the control spectrum B taken at the same KV (15), raster size and time period on none duct tissue. Following analysis, the carbon coated collecting duct tissue was coated with 3 nm Pt to verify the structural anatomy of the collecting ducts (seen here at lower magnification)

the data did show that some nanoparticles were retained in various tissues including the kidney, liver and brain. Schiffer et al. [44,45] also showed that radiolabeled QDs coated with polysorbate 80 rapidly dispersed to the heart, kidney and brain within the first 5 minutes post injection, whereas the same QDs without the surfactant coating were equally well tolerated by the mice, but the QDs did not travel to the brain within the 5-minute test period, suggesting that the coating of the nanoparticles with the polysorbate 80 in this case, altered the biodistribution of the nanoparticles.

The results with cellular cytotoxicity experiments showed that lung cells incubated with the QD diluted and QD stock solutions showed only a slight increase in cell necrosis compared to control values. However, when the stock solution dose was doubled to 2 μl, cell necrosis more than doubled. The colon cells seemed to be less sensitive to the QD preparations at this short time period, showing a small increase in cell death with short term exposure to the diluted and stock solutions but even less cell death following 1.5 hour exposure to the doubled concentration of the freshly prepared stock solution. These results suggested that short term exposure to these concentrations of old and freshly prepared QDs did not seem to produce acute cytotoxicity when the concentration of the QD dose was reduced and the time of exposure was brief. There was no difference in the responses of the lung and colon cells to the 'Old' and freshly 'New' stock solutions, suggesting that QD degradation had not occurred following refrigerated storage for eight months.

4.3 Shell Cross-Linked Nanoparticles (SCKs) with Rigid and Flexible Cores

Using a different type of nanoparticle with MicroPET analysis to study biodistribution in vivo, Sun et al. [51] found that nanoparticle size, core composition and surface PEGylation produced different results in nanoparticle retention and biocompatibility. Shell cross-linked nanoparticles (SCKs) are amphilphilic core-shell nanoparticles which achieve stability via cross-links throughout the shell layer of the micellar precursor. These SCKs were then further derivatized and in this case radiolabeled with Cu^{64} permitting in vivo biodistribution imaging by MicroPET to evaluate fate, blood retention and accumulation in liver, spleen, kidney and lungs. Evaluating the MicroPET biodistribution data from the PEGylated SCK nanoparticles compared to non-PEG derivatized parent nanoparticles in Balb/c mice showed the same low accumulation in blood throughout the experiment and no significant differences between the two at any given time period. They found that the more peptides they used to derivatize the SCK nanoparticles, the greater the serum IgG activity in the animals. Even though the presence of this antibody indicated an increasing recognition by the animals' immune systems that foreign material was in the blood, at all but the highest numbers of peptides, no gross abnormalities in organs were found postmortem. This fact indicates that lower peptide levels would be better to keep the animal's immune system from attacking the derivatized nanoparticles. Similar to the findings of Takeoka et al. [52] who used phospholipid vesicles, Sun et al. [51] also found that flexibility played an important part in biocompatibility and optimally designed a nanoparticle that would not become 'stuck' in vivo in organs. Nanoparticles having greater degrees of flexibility (flexible core materials) did not show high sequestration in organs. The larger flexible SCK nanoparticle did not get sequestered in lung, liver or spleen, whereas the smaller identically made particle accumulated comparatively more within these organs than its larger counterpart. The SCK nanoparticle with a rigid polystyrene core showed the longest blood retention and the lowest liver accumulation.

4.4 Summary

In summary, what these studies show about the interactions of these types of core-derivatized nanoparticles, both quantum dots (QDs), and shell cross-linked nanoparticles (SCKs) for use in in vivo biological studies is that:

- It is very important to consider the size of the particle to be engineered, as well as its flexibility and surface chemistry [51,52];
- The only way to know if a nanoparticle will behave as planned in vivo is to test it in the living organism to assess if the particle will stay in circulation, become sequestered in organs or tissues, or trigger an inflammatory or immune response;
- It is essential to collect data on cell and organism responses to engineered nanoparticles, and redesign the nanoparticle size, core, flexibility, and/or derivatization to optimize its biocompatibility for the desired application;
- Using lower doses of nanoparticles for shorter exposure periods, which have been coated with known biocompatible materials such as polysorbate 80 or PEG seems to favor in vivo biocompatibility; and
- Small physicochemical differences in nanoparticle design have very significant effects on biological cells in different parts of the body (brain, liver, lymph node, and spleen accumulation; blood inflammatory responses, clotting, antibody/immune responses), and therefore, one cannot assume that a nanoparticle that seems biocompatible in kidney tissue will also be universally biocompatible in blood, brain, reticuloendothelial tissue or any other organ or tissue in vivo.
- There are also inherent characteristics in each type of cell that determine whether a specific nanoparticle will be cytotoxic or biocompatible with that cell type.

5 Gold Nanoparticles

The use of gold particles for therapeutic applications for animals and humans is not new. For many years, intramuscular gold injections (i.e. Myochrysine, a gold sodium thiomalate) and more recently, orally administered gold preparations (Auranofin), have been used to reduce inflammation and pain for joint trauma and rheumatic diseases. The colloidal gold in these preparations distributes throughout the body (60% plasma protein bound), suppressing inflammatory and some immuno-arthritic responses in the joints of RA patients, and anti-arthritic disease modification in many mammals. The administration of commercial gold preparations depends on the ability of the recipient to tolerate the gold itself (some individuals react to gold metal), or increases in Au dose. Similar to the carbon nanoparticles and quantum dots previously discussed, the Au nanoparticles have a proclivity in vivo and in vitro to bioaccumulate within various types of cells with a special affinity for macrophage-type cells (both histiocytes and blood phagocytic cells), and reticuloendothelial cells throughout the body), these therapeutic gold preparations also produce varying degrees of gold bioaccumulation in such tissues as lymph nodes, bone marrow, spleen, adrenals, liver and kidneys [53].

Similar to the in vivo findings with quantum dots and carbon nanoparticles, Hillyer and Albrecht [54] found that following 4 days of colloidal gold (13 nm diameter nanoparticles) administration (by intraperitoneal injection) to BALB/c mice, the tissues analyzed by instrumental neutron activation analysis (INAA) showed the highest gold deposition in the spleen and liver. Moderate localization in the lung, heart, kidney, stomach and intestine was also seen, with the lowest Au content found in the brain [54]. Verification of Au deposits in TEM sections by x-ray microanalysis revealed that Au nanoparticles in liver were localized to the Kupffer cells. In the spleen, Au nanoparticles were found in macrophages of the marginal zone surrounding the white and red pulp. In the kidney, gold deposition was in three places: the intraglomerular mesangial cells; within the filtration phagocytic cells; and within the urinary space. The Au nanoparticles were also observed in membrane bound vesicles within the proximal and distal convoluted tubules, which indicate specific cellular uptake and transport by these cells. They [54] observed that in all of these organs and tissues, the Au nanoparticles were seen as aggregates and as single nanoparticles either free in the urine or blood, or contained within membrane bound vesicles or phagocytic vacuoles. Animals sacrificed within 3 hours of the last Au treatment showed Au nanoparticles within blood monocytes, as well as Au nanoparticles in membrane bound vesicles within the endothelial cells lining the blood vessels. This indicates that there may be commonalities between Au nanoparticle biodistribution and cellular localization with other types of nanoparticles, when the nanoparticle itself does not produce immediate cytotoxicity.

These gold studies, and the previous quantum dot studies did not look at long term in vivo survival of the treated animals, and there is no assurance that the cells with Au nanoparticle accumulated in vacuoles and vesicles will remain static throughout the life of the animals. However, if Kupffer cells in the liver and macrophages in the spleen, lymph nodes, blood and major organs are engorged with nanoparticles, those tissues may not be able to remove toxins, foreign particles and microorganisms from the body as efficiently as in normal tissue. Therefore, excessive bioaccumulation may pose a threat to the organism in terms of survival, even when the bioaccumulated nanoparticles are benign, because the cells designed for removal and destruction of foreign materials and toxins, or disease causing organisms become too engorged to function properly (causing problems with cell division and protein synthesis).

The successful use of gold particulates in veterinary and human therapeutic treatments suggests some degree of biocompatibility. This has encouraged the development of new forms and modifications of Au nanoparticles for biomedical, biological and treatment/diagnostic applications, as well as the use of these biofriendly nanoparticles in self assembly paradigms using biomolecules such as DNA, antibodies, various enzymes and other proteins [55–57]. This section will present experimental data using citrate capped and DNA-functionalized Au nanoparticles with human cells [5], to evaluate binding uptake and cytotoxicity following exposure.

5.1 Gold Nanoparticle Responses with Specific Cells In Vitro

Li et al. [58] recently reported that incubation of gold nanoparticles (0.5 nM and 1.0 nM) with MRC-5 embryonic lung fibroblasts produced oxidative damage to the cells in vitro, and caused down regulated genetic expression. By TEM, they found that incorporated gold nanoparticles were seen within membrane limited cytoplasmic vesicles, as well as within the cytoplasm itself, indicating deterioration of the vesicular membrane. In 2005, Tsoli et al. [59] demonstrated that 1 nm diameter spherical Au nanoparticles passed into cells and into their nuclei, with the Au nanoparticles attaching to the nuclear DNA. Chithrani et al. [60] examined the interactions of spherical and rod shaped colloidal gold nanoparticles (ranging in size from 14–100 nm) with Hela cells in vitro. Following 6-hour incubation with various gold particles, they found nanoparticles trapped within Hela cell cytoplasmic vesicles, with a maximum nanoparticle size of 50 nm [60]. Round Au nanoparticles (14 and 74 nm spheres) were incorporated into cells preferentially to rod shaped (74×14 nm) particles. When citric acid stabilized Au nanoparticles were tested, the overall surface charge of the nanoparticles was negative, and the citric acid was weakly bound to the nanoparticle. They postulate that the citric acid could be desorbed from the metal surfaces by proteins (like the serum proteins in the media), and that this process appeared to be instantaneous. These serum proteins on the surface of the Au nanoparticles were felt to dictate nanoparticle uptake using multiple surface receptor-mediated endocytosis. This suggested that in the Hela cell, the Au nanoparticles could be transported into the cells in larger numbers if the nanoparticles were within a specific size range, had a spherical shape, and were allowed to self-adsorb serum proteins (rather than be coated with a specific protein'transferrin'), to permit multiple receptor-endocytic transport. In no cases did Chithrani et al. [60] report that the nanoparticles entered the nuclei of the cells within the 6-hour incubation period. Pernodet et al. [60] studied the dose/concentration responses of human dermal fibroblasts in vitro with citrate capped 14 nm Au nanoparticles. They found that these nanoparticles easily crossed the cell membrane and accumulated in vacuoles. After 6 days of incubation, the fibroblasts were engorged with vacuoles containing Au nanoparticles, and the cells began to show signs thereafter of alterations in normal actin structure and metabolism, as well as changes in extracellular matrix (ECM) structure. All of these studies suggest that gold nanoparticles can produce stress on cells in vitro, as well as alter genetic expression even when the gold nanoparticles do not enter cell nuclei.

Once again, it seems that gold nanoparticles follow the same general "laws of nanoparticle engagement" with biological cells that we have previously discussed. The size, outer surface characteristics (reactive groups, morphology, charge, exposure of the crystalline lattice) and functionalization profoundly affect whether an Au nanoparticle will interact in a biocompatible or destructive way with a living cell. These studies have raised several important points that have been discussed earlier in this lecture in regard to other types of nanoparticles.

- The size and shape of the nanoparticle seems to matter in terms of cell uptake and localization;
- Smaller nanoparticle size does not always assure maximum cellular incorporation or cytotoxicity. Specific types of cells have a "preferred nanoparticle size or shape" for facilitated uptake;
- In biological systems, nanoparticles may spontaneously change their outer surface characteristics via 'self-functionalization' with biomolecules- which, in

turn, affects cell uptake, intracellular processing and bioaccumulation or degradation/excretion- in other words, the fate of the nanoparticle;

- Nanoparticle incorporation into a cell does not necessarily mean that the nanoparticle will enter organelles. Specific criteria apparently must be met before a nanoparticle is capable of entering organelles (i.e. mitochondria, nuclei, nucleoli, Golgi); and
- Nanoparticle entry into cells and storage in vacuoles or endocytic vesicles does not necessarily cause cytotoxicity, but can overload the cell's ability to synthesize proteins, exhibit normal motility, divide normally and maintain normal cytoskeletal / actin function.

6 Methods and Experimental Results

Gold Nanoparticle Synthesis and Characterization: We have been studying the interactions of different types of Au nanoparticles with human epithelial cells (lung NCI-H292 cells and Caco2 colon cells). As an engineered nanoparticle, Au nanoparticles are especially adaptable . They can be made in various sizes and functionalized in many ways to suit myriads of applications. In collaboration with Drs. Oleg Gang, Daniel van der Lelie, Mathew Maye and Dmytro Nykypanchuk, we have been examining how human epithelial cells bind and incorporate citrate capped and DNA-functionalized 10 nm and 2 nm Au nanoparticles, and whether these nanoparticles induce cytotoxicity [5].

Synthesized citrate capped 10–13 nm diameter (30 nM, 0.18 mg/ml) and 2 nm diameter (6.2 nM, 1.01 mg/ml) Au nanoparticles were incubated with human colon and lung cells in vitro. Au nanoparticles (10–13 nm diam.) were also functionalized with single stranded DNA (ssDNA), so that the gold nanoparticle core was 20% ssDNA, or 100% ssDNA, covered. The nanoparticles were analyzed in solution to measure size, distribution, uniformity, morphology, volume and mass using transmission electron microscopy, dynamical light scattering/Zeta potential analysis and UV-Vis. Citrate capped Au nanoparticles were resuspended in 0.3 M PBS to make a final dilution of 0.03 μM (0.2 mg/ml) for the ∼10 nm diameter citrate capped Au nanoparticles and 0.1 μM (0.5 mg/ml) for the 2 nm diameter citrate capped Au nanoparticles, which permitted a final pH of 6.5–6.8, which was compatible with the human cell preferred hydrion concentration. A color indicator in the cell culture medium indicated if the culture medium was at the optimum pH for cell growth.

The 10 nm core Au nanoparticles were functionalized with short ssDNA oligomers of 30 nucleotides with a 3'thiol modification described by Maye et al. [55], producing a nanoparticle with coiled ssDNA wrapped around the central Au-core.

Using methods for high coverage (∼50ssDNA per Au core), Maye et al. [55] were able to label the gold core with DNA essentially covering it (100% coverage seen in Fig. 16. The nanoparticles were stable in high salt concentrations, and therefore could tolerate the conditions during cell incubation. The 100% covered DNA-Au nanoparticles were prepared in 0.3 M PBS in a concentration of 125 nM

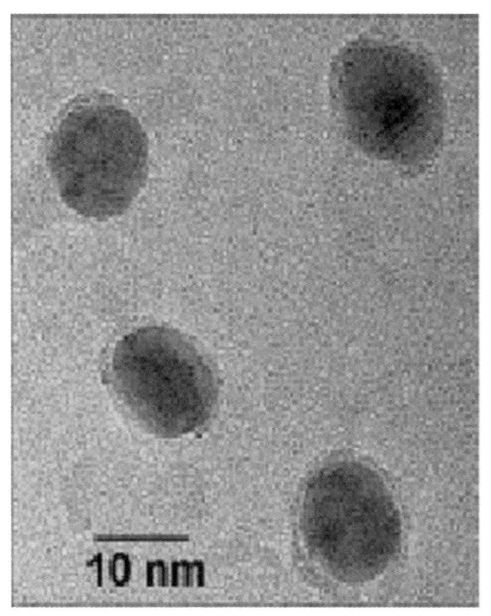

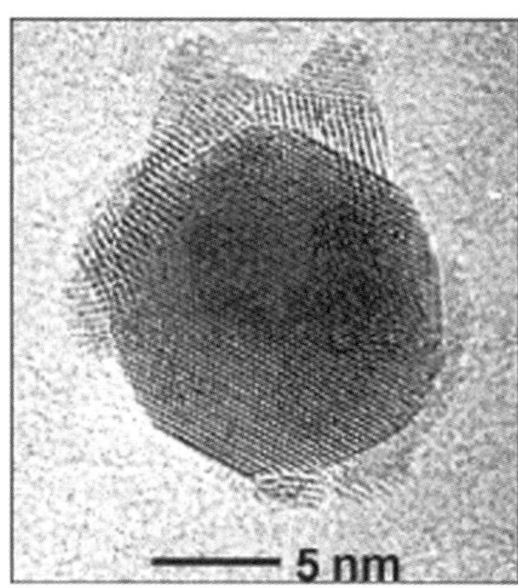

Fig. 16 TEM images of uranyl acetate stained 100% ssDNA covered Au nanoparticles. The halo around the gold core is the lightly stained DNA [55]

(0.75 mg/ml). The 20% DNA covered Au nanoparticles enabled us to study the biological interactions to the completely covered (100% DNA covered) nanoparticle as compared to a nanoparticle of the same size and composition but with only 20% ss-DNA (150 nM, 0.95 mg/ml) and 80% of the gold core surface exposed to the living cells. All of the DNA-Au nanoparticles were diluted in PBS to a concentration of 13–14 μM (0.9 mg/ml). Following monolayer growth to 94–95% confluency, 1 μl dose of the citrate capped or DNA-Au nanoparticles was added to 1.5 ml of cell culture medium in each shell vial.

Cell Culture Experimental Methods: Following 1.5 or 3 hour incubation at 36°C with either the citrate capped or DNA-functionalized nanoparticles (1 μl dose/monolayer) with constant swirling, the media was removed from the cell culture vials (for UV-Vis analysis and/or microscopic analysis to screen for nanoparticles, contaminants and dead cells). The cell monolayers were washed two times in warm PBS, and the cells prepared for light microscopy vital staining (and photography), transmission electron microscopy or field emission scanning electron microscopy (as described previously). Control monolayers were given 1 μl of PBS and identically incubated and analyzed.

Viability Testing: Colon cells incubated with citrate capped- or DNA functionalized 10 nm Au nanoparticles for 1.5 hours showed no significant increase in cell necrosis compared to control values. At 3-hour incubation, there seemed to be only a slight increase in cell death compared to control values, even when the dose of the gold nanoparticles was doubled (Fig. 17). Lung cells similarly showed very little cytotoxicity to the Au nanoparticles. However, when 2 nm citrate capped Au

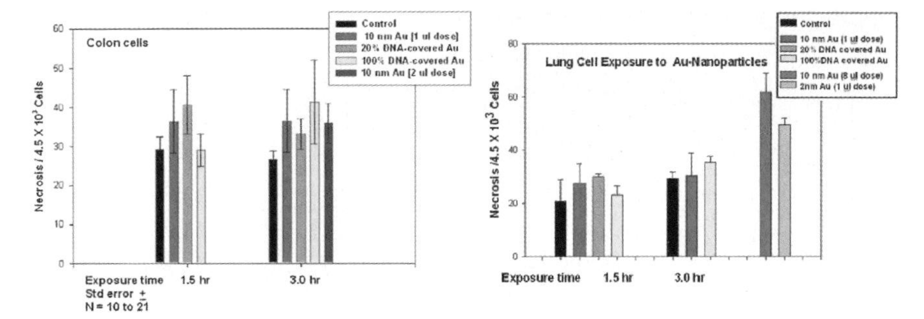

Fig. 17 Human Caco-2 colon cells (*left*) and human lung NCI-H292 cells(*right*) incubated with 10 nm citrate capped Au nanoparticles, or incubated with 20% - or 100%-DNA covered 10 nm Au nanoparticles for 1.5 or 3 hours. In general, even when the 10 nm citrate capped Au nanoparticle dose was doubled, the colon cells did not show increased necrosis compared to control values. The lung cells similarly did not show necrosis levels above control values unless they were incubated with far higher doses of the citrate capped 10 nm Au nanoparticles or the 2 nm citrate capped Au nanoparticles for 3 hours

nanoparticles were incubated for 3 hours with lung cells, cytotoxicity was seen (not shown is 1.5-hour 2 nm citrate capped viability data which also was in the control viability range, but not entered here because only 3 monolayers were counted compared to the 10–21 monolayers used for the compilation of data for the other histogram bars). When the dose of the 10 nm citrate capped Au nanoparticles was increased from a 1 μl dose to an 8 μl dose, it too produced an increase in necrosis at 3-hour incubation (Fig. 17). From these data, it would seem that only the citrate capped nanoparticles seemed to be cytotoxic lung cells, with none of the Au nanoparticles producing severe necrosis to the colon cells. Further studies are in progress to see if the viability of the lung and colon cells is altered when the dose of the 100% DNA- and 20% DNA covered Au nanoparticles are also doubled. **After screening identical samples by TEM and FESEM, we did not see the same cell normalcy (in terms of ultrastructure and intracellular processing of nanoparticles, Figs. 18–20), suggested by this type of vital staining for necrotic cells, and we realized how important it was to use multiple methods of analysis for determining biocompatibility.**

Microscopy Results

Citrate capped Au nanoparticles: FESEM imaging of colon and lung cells incubated with 10 nm and 2 nm citrate capped Au nanoparticles showed aggregations of nanoparticles on the surfaces of the tissue culture cells at 1.5 hour incubation with little damage to the apical cell membranes although a great deal of surface membrane activity was visible on many cells (Fig. 18). Membrane blebbing, membrane ruffles and small filiform apical membrane projections were all seen on colon and lung cells (Fig. 18). In addition, lung cells with Au nanoparticles clustered on

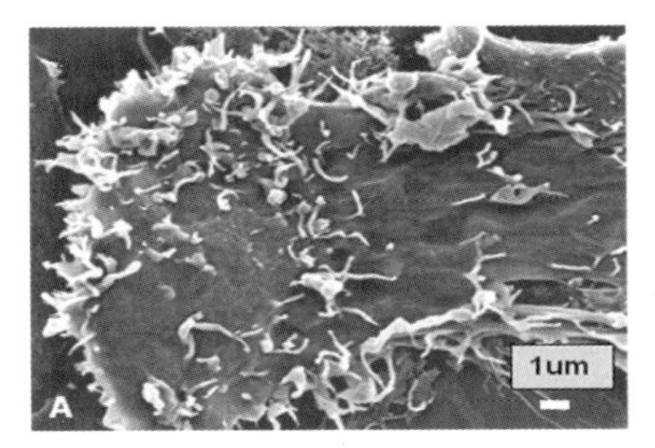

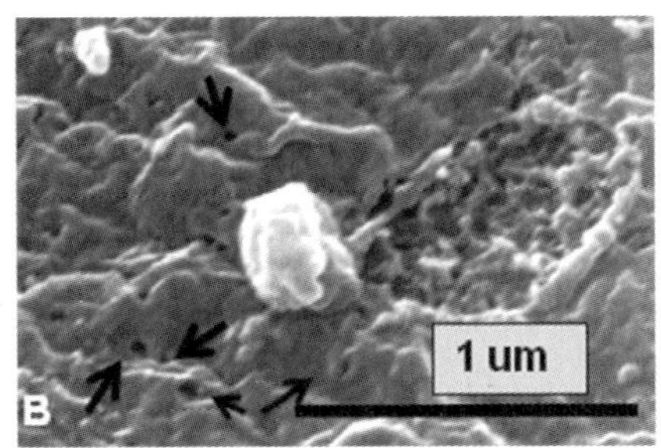

Fig. 18 Lung cells incubated with 10 nm citrate capped Au nanoparticles showed blebbing and membrane ruffling following 1.5–2 h exposure (Fig. A). At 3 hour exposure (Fig. B), the membrane surface revealed small, punctuate holes (*arrows*) in the plasma membrane, as well as larger patches of missing plasma membrane

the apical surface also exhibited large cytoskeletal projections of the apical plasma membrane (Fig. 19) [5]. At 2 and 3 hour incubation with the same nanoparticles, cells appeared to have small holes in the apical plasma membrane (Fig. 18), which were not seen in any of the control samples. This was especially true of cells incubated with the 10 nm citrate capped Au nanoparticles, where small patches of apical membrane were missing (Fig. 18, arrows). Unlike the previously shown carbon nanotube interactions with lung and colon cell apical plasma membranes (showing tears and destruction of the surface membranes), the holes in the membranes following gold exposure were not tears, and they did not expose the underlying cytoplasm. The cells remained attached to the coverslips and to one another, suggestive of normal cell-cell adhesion and cell-substrate adhesion.

TEM of identical preparations revealed gaps or holes in the apical membrane with Au nanoparticles passing into the cell cytoplasm at these sites [5]. Lung and colon cells exposed to the 10 nm citrate capped Au nanoparticles appeared plump and intact by FESEM, however by TEM, Au nanoparticles were seen within large intracellular vacuoles not typical of these cells. The membranes lining the vacuoles were torn or partially missing in those areas containing Au nanoparticles, suggesting that lipid peroxidation or ROS focal damage of the internal cellular membranes had occurred [5]. Vacuoles were also observed within nuclei, which were never seen in the Controls or in other nanoparticle-treated cells. Individual cells containing single or clustered 10 nm Au nanoparticles often lacked intact vacuolar and organellar membranes. Membranes of the nuclei, mitochondria and endoplasmic reticulum were not visible within the cells, however the cells remained attached to the substrate and did not show signs of apoptosis. Often nanoparticles were seen in small membrane-limited vesicles in the central and basal cytoplasm of the cells, suggesting endosomal transport of the nanoparticles. The destruction within these cells was severe, yet by viability testing (vital staining), the cells incubated with 10 nm citrate capped nanoparticles showed no significant necrosis compared to control values, even after 3-hour incubation. FESEM did not show increases in apoptotic or

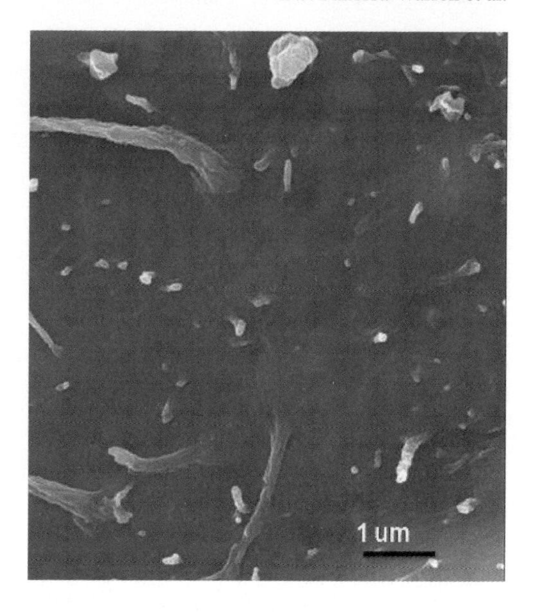

Fig. 19 FESEM of the surface of a lung cell following 1.5 h incubation with 2 nm citrate capped Au nanoparticles showed numerous cytoskeletal projections of varying lengths on the cell surface, and an intact apical membrane.

detached cells, yet by TEM, these cells were highly vacuolated and had damaged intracellular architecture and nuclei.

2 nm Citrate Capped Au nanoparticles: Cells incubated with 2 nm citrate capped Au nanoparticles appeared normal by FESEM. Lung cells showed cytoskeletal extensions of the apical cell surface similar to those seen with the 10 nm Au nanoparticles when bacteria or particulates were present in the cell culture. By TEM, aggregations of Au nanoparticles were seen attached to extracellular matrix material. These clusters of nanoparticles embedded in cell-elaborated apical surface protein or mucous containing material remained attached to the apical surface membrane or to cytoskeletal projections on the apical cell surface even after tissue processing (Fig. 19).

The cytoskeletal projections of the lung apical surface seen by FESEM in Fig. 19 were also seen to be continuous extensions of the apical surface by TEM (Fig. 20). The 2 nm citrate capped Au nanoparticles were never seen as individual nanoparticles, but were always embedded in adsorbed protein/mucus typically produced on the lung cell surface. These aggregations (seen in Fig. 20A, arrows) passed into the apical part of the cell. We did not see traditional endocytosis incorporation. Instead, the nanoparticles seemed to pass into small holes in the plasma membrane that appeared at the time that the Au nanoparticles touched the membrane surface. However, once within these holes or pits on the cell surface, the nanoparticle aggregates became enclosed within vesicular membranes and were transported to larger vacuoles and vesicles (V) which also contained membrane fragments (Fig. 20A). These could represent small fragments of plasma membrane that had been attached to the nanoparticles during entry, and were now undergoing membrane recycling within the cytoplasmic vesicles. The proteins and membrane components stripped off in the vesicles could then be recycled to make new membrane. This seems similar to a

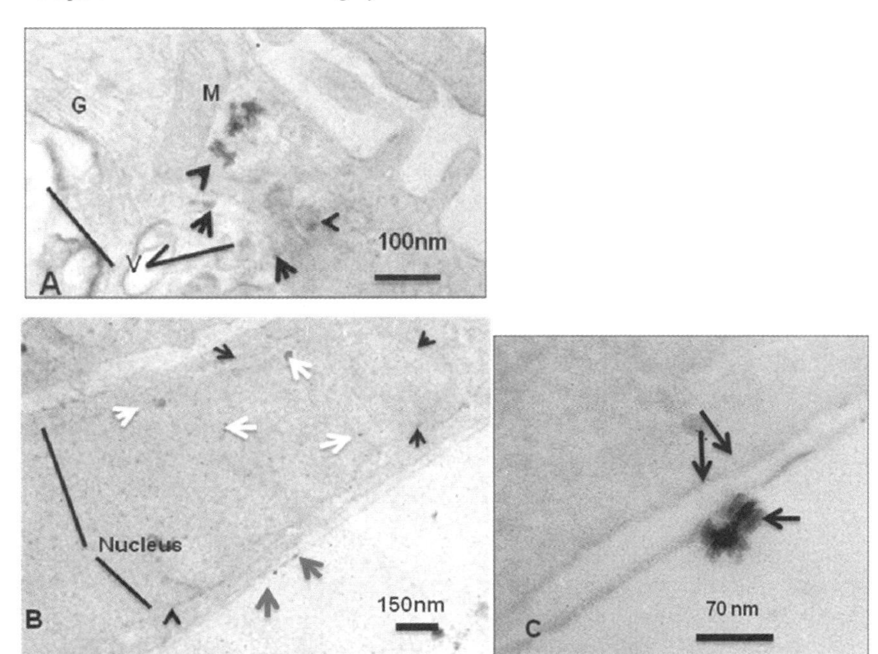

Fig. 20 TEM of lung epithelial cells incubated with 2 nm citrate capped Au nanoparticles. (A). Cytoskeletal projections can be seen on the apical surface at 1.5 hour incubation (Fig. A). *Arrows* point to gold aggregates in the apical region. (M) Mitochondrion, (G) Golgi apparatus, and (V) intracellular vacuoles. Figure B shows a nucleus containing Au nanoparticle aggregates (*white arrows*) within the nucleoplasm at 3 hour incubation. The nuclear membrane is intact (*black arrows*), and the cell shows no signs of vacuolization or cell stress, but the nucleus lacks heterochromatin. Excreted Au nanoparticle aggregates (Fig. B, *larger arrows*) can be seen at the basal membrane of the cell. Figure C. At high magnification, the Au nanoparticle aggregates (*arrow*) can be Fig. B clearly seen excreted from the outside of the basal cell membranes. The double arrows show the intact membrane of the basal infoldings

type of nanoparticle/membrane recycling we saw and reported [4,5] in earlier studies involving carbon nanoloops binding to gClqR-invasion proteins on human colon and lung cells during microbial attack.

No Au nanoparticles were seen within ER, mitochondria (M) or Golgi (G). After 1.5-hour incubation, Au nanoparticle clusters were seen within cell nuclei (Fig. 20B, white arrows) of viable cells. Unlike the control cells that had darkly staining heterochromatin in all cell nuclei, the 2 nm citrate capped Au nanoparticle incubated cells revealed nuclei with intact nuclear membranes, but no heterochromatin. Although the 3 hour incubated cells seemed viable and had normal mitochondria, endoplasmic reticulum, Golgi and other ultrastructurally normal appearing intracellular structures, their nuclei lacked visible heterochromatin (the site of condensed DNA in the nucleus) [5]. Theoretically, 2 nm nanoparticles could enter a human lung or colon cell nucleus through the nuclear membrane pores [5]. If the 2 nm gold aggregates adsorbed to protein or mucoid material remained flexible while in the cell, they could pass through the 6–9 nm pore opening in the nuclear membrane and reside in the nucleoplasm.

These 2 nm citrate capped Au nanoparticles may have their citrate cap removed at the cell surface, and rapidly replaced with glycoproteins from the apical surface. The 2 nm Au nanoparticle core would be highly reactive without the citrate cap, and may become embedded in mucoid or extracellular matrix macromolecules produced on the apical surface during incubation. Either, the Au nanoparticles can enter the apical surface because they are coated with the cells' own apical surface macromolecular elaborations and the cells will take this material in for cellular recycling or perhaps some areas of the Au-core may be exposed to the apical plasmalemma, which could cause small focal areas of damage to the outer lipid layer of the plasmalemma (by lipid peroxidation) to gain entry into the cell through small holes at the sites of nanoparticle-aggregate attachment to the membrane. This process might not be initially lethal to the cell because the cell membrane is predominately intact. Within the cell cytoplasm, the gold core may again adsorb protein and reside in a vacuole or vesicle until it is processed and excreted at the basal surface, or until the adsorbed protein on the Au-core surface is recycled and re-used by the cell. This disruption of the vesicular membranes would permit the Au nanoparticles to escape their vesicles and move around the cell without the benefit of endocytosis, or actin-driven intracellular transport, allowing them to travel to the nuclei. Since there is a continual stream of messenger molecules being passed into the nucleus and entering through the nuclear pores, Au nanoparticles attached to appropriate intracellular RNA or proteins may enter the nucleus during normal nuclear-cytoplasmic communication, but become trapped. Once the Au nanoparticles form sufficiently large aggregates (greater than 9–12 nm diam.), they can no longer easily pass through the nuclear pores.

DNA-functionalized Au nanoparticles: In some earlier work, we found that 20% ssDNA- covered 10 nm Au nanoparticles produced holes in the apical membranes of both colon and lung cells in vitro [5]. Similarly, the 20% and 100% ssDNA covered Au nanoparticles also produced holes in the apical surface of cells in vitro (Figs. 21 and 22).

During the 3-hour incubation period, these cells did not detach from the coverslip surface, nor show signs of acute cytotoxicity or distress. The 20% DNA covered Au nanoparticles produced no apical surface damage at 1.5-hour incubation, but as aggregates of nanoparticles accumulated on the apical surface, myriads of holes were seen in the plasma membranes of the cells (Fig. 21). Individual Au nanoparticles could be seen in some holes, but the surfaces of the cells revealed increasing damage to the membranes after 3-hour incubation. The 100% DNA-covered nanoparticles seemed more benign to the cells producing smaller holes in the surface layer of the cells (Fig. 22) and a rapid novel form of transport through the cells, with excretion of the gold at the basal surface. By TEM, the 100% DNA-covered nanoparticles seen on the apical cell surface (Fig. 23A), appeared to pass through the plasmalemma without visible endocytic membranes. Nanoparticles were seen within tubular channels very similar in morphology to endoplasmic reticulum (ER) (Fig. 2 B). Gold nanoparticles appeared in these channels not as discrete 13–15 nm spheres but as electron dense aggregations which passed from the ER tubular channels to the

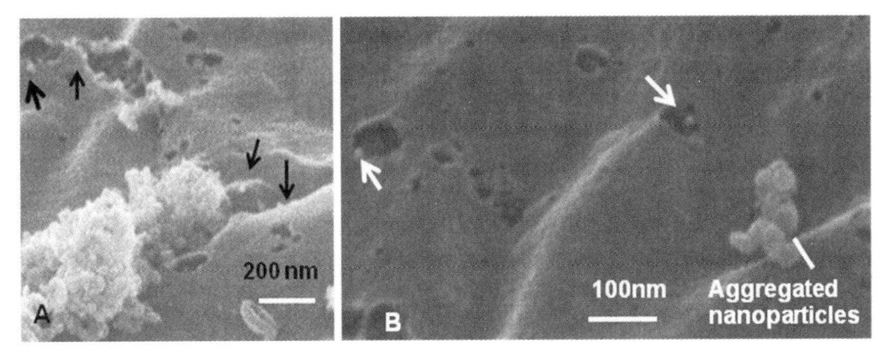

Fig. 21 20% ssDNA covered Au nanoparticle aggregates enmeshed in cell surface protein or mucus, or individual DNA-Au nanoparticles (*arrows*) were seen on the apical plasma membrane (Fig. A). Individual holes in the plasma membrane were associated with the sites of attached nanoparticles, here seen as a larger aggregate (Fig. B), or as single nanoparticles within the membrane holes (*white arrows*). With the Pt coating, individual DNA-Au nanoparticles averaged 18–24 nm diameter

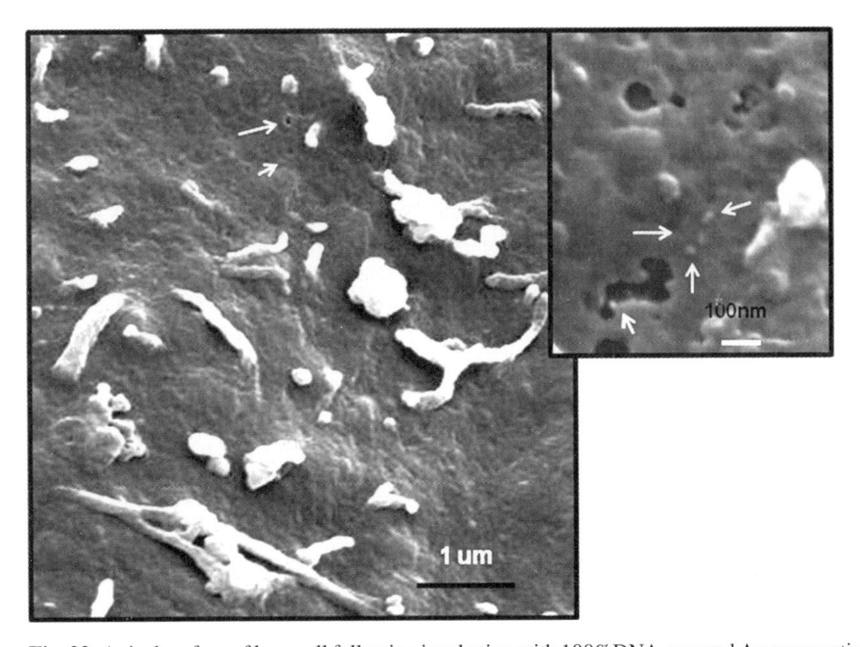

Fig. 22 Apical surface of lung cell following incubation with 100%DNA-covered Au nanoparticles showed typical cytoskeletal filiform extensions on the cell surface and some openings (*arrows*) in the plasma membrane. The high magnification (insert) shows joined double 100% DNA covered Au nanoparticles on the cell surface (*arrows*), and one Au nanoparticle within an opening in the plasma membrane surface (*white arrow*)

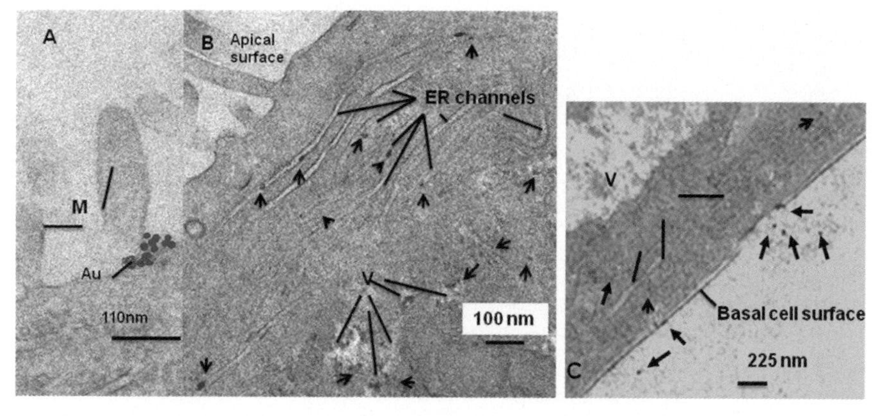

Fig. 23 100% ssDNA covered Au nanoparticles (Au) on the apical colon cell surface were seen adjacent to the microvilli (M) on the apical plasma membrane (Fig. A). After 1.5 hours (Fig. B), the nanoparticles (*arrows*) entered the apical cell surface and passed into intracellular channels (ER) and were seen within central (V) vacuoles (Fig. B) and later in basal vacuoles at 3-hour incubation Fig. C). Numerous channels parallel to the basal cell surface (lines point to channels, Fig. C) contained nanodots, with aggregates excreted at the basal surface (*arrows*, Fig. C)

Golgi apparatus [5] and into vacuoles located centrally in the cell (Fig. 23B). The Au nanoparticles could be seen as discrete electron dense dots in the lower basal vacuoles of the cell. Discrete electron dense nanodots were seen passing into another extensive set of tubular channels parallel to the basal membrane of the cells (Fig. 23C). Nanoparticles were excreted at the basal plasma membrane.

The highly polarized colon epithelial cells used in this study usually have a basal surface with numerous basal infoldings of the membrane surface. Following 100% DNA covered Au nanoparticle exposure, the basal infoldings were gone and had been replaced by these arrays of ER-like tubular channels containing Au nanoparticle core fragments. The aggregates of gold dots that were ejected from the cells at the basal membrane (Fig. 23C), were never seen to build up within the cells in vacuoles, endosome or became associated with organelles such as the nucleus or mitochondria within the 3-hour incubation period. The incorporation and transport of the DNA-covered Au nanoparticles did not follow usual cellular mechanisms for particulate incorporation and intracellular processing in these epithelial cells. This unique type of nanoparticle cellular processing is summarized in the schematic Fig. 24. For comparison, Fig. 25 shows the more typical type of cellular incorporation and intracellular transport of particulates, macromolecules and microorganisms seen in epithelial and other types of cells, called **endocytosis**.

Summary Au nanoparticles

Citrate capped Au nanoparticles entered both lung and colon cells within 1.5 hour exposure. Cells with incorporated 10–13 nm diameter Au nanoparticles seemed

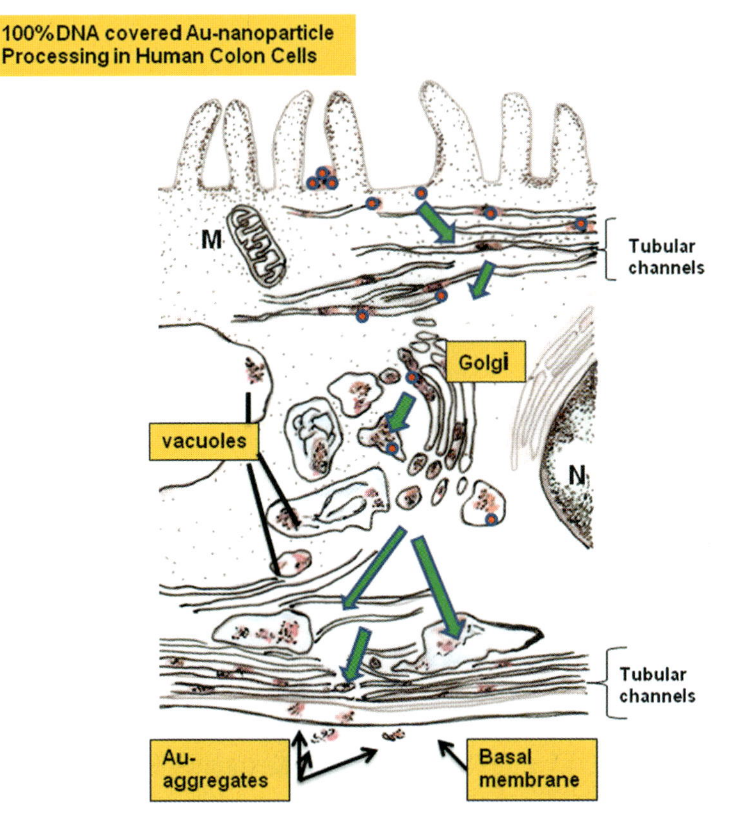

Fig. 24 Schematic diagram of 100%DNA covered Au nanoparticle incorporation and transport in colon cells. DNA covered nanoparticles (*red dots*) entered at the apical surface and passed into tubular channels similar in appearance to ER. They traveled through these channels to the Golgi apparatus and into vacuoles where they no longer had the distinct 10 nm diameter core but appeared as electron-dense fragments of nanoparticles. Within vacuoles, these fragments became aggregates which passed into tubular channels at the basal portion of the cell. Clusters of aggregates were excreted at the basal membrane. No Au nanoaggregates or 10 nm nanoparticles were seen in the nuclei (N) or mitochondria (M) by TEM

to show no severe necrosis even after 3-hour incubation at the 1 μl and 2 μl doses. However, by TEM, these cells showed serious intracellular vacuolization and damage (dissolution and tears) in intracellular membranes including those of the mitochondria, Golgi and nuclei. The 2 nm Au nanoparticles showed no production of abnormal intracellular vacuolization, nor any damage intracellularly to mitochondria, Golgi or intracellular membranes. However, the nuclei of the lung cells incubated with 2 nm Au nanoparticles for 3 hours showed nanoparticle aggregates within the nuclei and no nuclear heterochromatin: both of these findings are not compatible with normal cell function. Viability analysis of lung cells exposed to 2 nm citrate capped Au nanoparticles showed elevated necrosis compared to control values, and compared to the DNA-covered Au nanoparticles.

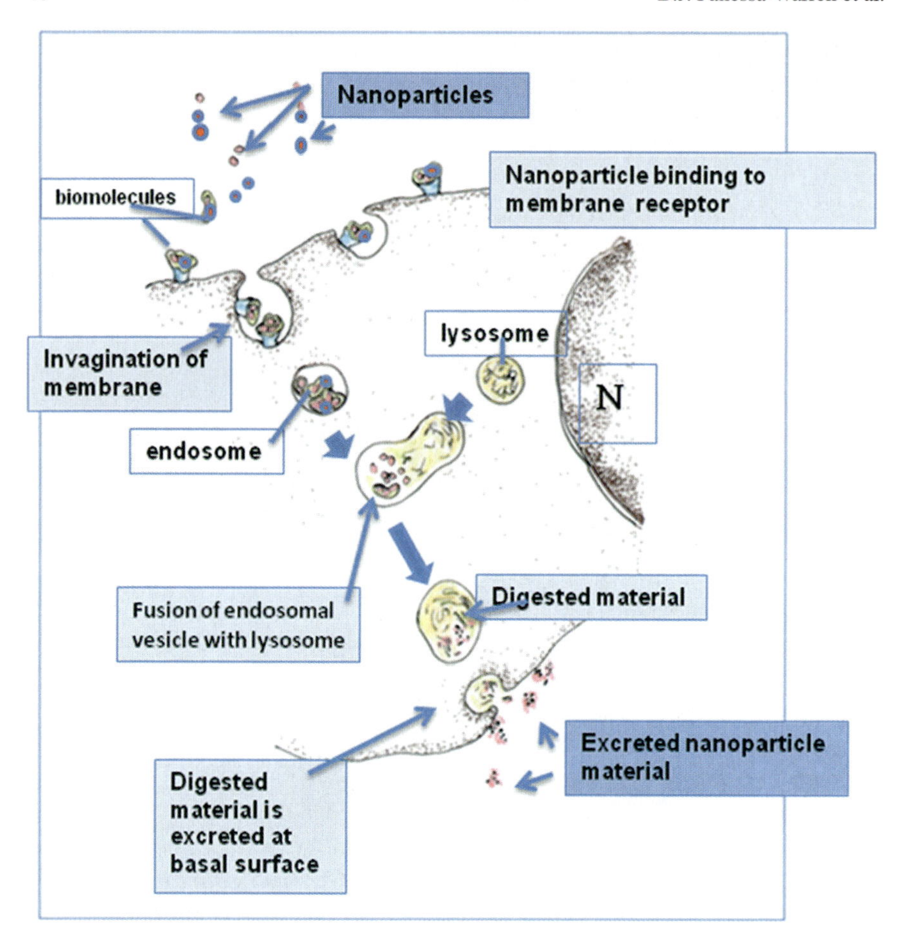

Fig. 25 Endocytosis. Typically non-phagocytic cells will incorporate microorganisms, biomolecules and foreign material by a process called endocytosis. Traditionally, materials attach to receptor sites (*in blue*) on the apical plasma membrane. In this diagram, the red nanoparticles adsorb biomolecules (*in green*) such as serum proteins or surface glycoproteins, which then bind to the receptor sites on the membrane. The plasma membrane with the attached nanoparticle-receptor complexes invaginates to form the endosomal vesicle. The nanoparticles, membrane receptor and adsorbed protein are contained within the membrane bound vesicle and are not free within the cytoplasm. The vesicle can then break free of the apical surface and travel in the cytoplasm to the basal surface of the cell

The DNA-covered Au nanoparticles showed much better biocompatibility (viability and TEM analysis) than the citrate covered nanoparticles. The unusual finding of holes in the plasmalemma of colon and lung cells incubated with these nanoparticles, and possible entry and processing of the nanoparticles by some modified pathway leave many remaining questions. Whether the entry of these DNA-Au nanoparticulates into the cell surface was achieved through focal membrane damage, or a different form of transport mediated by the DNA coating is not clear without further research. Neither coated nor uncoated endosomal vesicles (Fig. 25) were evident in any of the TEM images of the apical cell cytoplasm. Cellular Processing of the nanoparticle may have been determined by the DNA coating, in much the same way that microbial DNA is incorporated, processed and neutralized by non-immune and non-phagocytic biological cells.

Endocytosis versus altered transport: During phagocytosis and endocytosis, lysosomes produced by the cell travel to the vacuoles or vesicles containing foreign material and fuse with the membrane of the phagosomes or endosomes, thereby initiating enzymatic digestion and recycling of entrapped materials (microorganisms, proteins, particulates, membranes, etc.). The undigested materials are transported to the basal surface of the cell, where the endocytic vesicle membrane fuses with the cell's basal plasmalemma, and non-digested material is excreted. This allows the undigested and unused materials to be excreted from the cell. In the case of professional phagocytes, materials in phagocytic vacuoles can be retained for extended periods of time without successful digestion or excretion [62].

Of significance in this process is that the nanoparticles bound to receptors on the membrane surface during the process of endocytosis never truly cross the plasma membrane [14]. They are enveloped within a vesicle formed in part by the plasma membrane [14] which keeps the contents of the endosomal vesicle apart from the cell cytoplasm and all organelles within the cell. Therefore, nanoparticles do not cross the plasma membrane and enter the cell cytoplasm under normal conditions. If a nanoparticle generates ROS and induces lipid peroxidation of the endosomal or phagocytic membranes during transport within these vesicles and vacuoles, it can destroy this protective barrier and exit the vesicular membrane. Under these abnormal conditions, nanoparticles can enter the cell cytoplasm and cause further damage to the cytoplasmic constituents and intracellular membranes, as was seen with the 10 nm citrate capped Au nanoparticles. This occurrence is usually the harbinger of the future demise of the cell. However, if nanoparticles are used as carriers for proteins or DNA which is meant to be removed by the cell from a biocompatible nanoparticle during endocytosis or phagocytosis (during lysosomal digestion and recycling), the resultant protein or DNA can be released into the cell's cytoplasm, and the nanoparticles excreted through normal endocytic excretion processing.

7 Conclusions

Currently, there appears to be conflicting results from various investigators on whether these types of carbon nanoparticles, quantum dots, SCK nanoparticles and gold nanoparticles are cytotoxic or biocompatible. There does not seem to be any general mechanism for making these nanoparticles universally 'non-toxic' to all living cells and all organisms. However, there are important findings that can be applied for increasing nanoparticle biocompatibility and reducing cytotoxic interactions in vivo and in vitro. Some of these are:

- The bare surface area of many nanoparticles seems to be highly reactive with biological cells and biomolecules. Consequently, exposing these reactive nanoparticle surfaces usually increases toxicity –but, in each case, this must be experimentally verified.
- Using the lowest nanoparticle dose to get the desired response for the shortest period of time, in general, seems to promote biocompatibility.
- Nanoparticles that have been sufficiently cleaned to remove catalyst (i.e. metals) and other debris associated with synthesis appear to produce less cytotoxic effects.
- Coating a nanoparticle can make it more biocompatible if the outer coating completely covers the nanoparticle reactive surface, and-

    1. Cannot be removed and utilized by the living cell, leaving the nanoparticle surface exposed to cytoplasm and intracellular membranes;
    2. It has a uniform continuous covering, lacking any cracks, roughness or interruptions that could lead to complement or antibody attachment, or dissolution of the coating by cell digestion.

- It is important to test nanoparticle/biological interactions experimentally and modify the nanoparticles for best biocompatibility with the cells, or the route of entry to be used in order to eliminate membrane lipid peroxidation; reduce the generation of reactive oxygen species; prevent acute and chronic release of inflammatory factors (and 'complement' activation); guard against alterations in genetic cellular function; and reduce the possibility of nanoparticles becoming 'stuck' during filtration or passage through pores and fenestrations [51] due to size, inflexibility of the nanoparticle core, or protein adsorption and agglomeration.
- Once nanoparticles are within tissues of an organism, they may translocate to new areas via cellular or fluid transport and accumulate in new areas that could challenge cell viability and normal function of the cell, tissue or organism.
- Nanoparticles that are highly oxidized, or have very reactive surfaces following cleaning/cutting or functionalization, usually cause acute and/or chronic cytotoxic responses the longer the nanoparticle is in contact with the cells and tissues.
- Different cells show different responses to nanoparticles based on characteristics inherent in their membrane surface reactivity and cellular genetic programming.
- Since nanoparticles and especially carbon nanoparticles can selectively bind blood proteins and cell-elaborated biomolecules, model biological systems need to be developed and used to test these nanoparticles under various conditions to insure that unforeseen biomolecular binding at the nanoparticle surface does not change nanoparticle function in vitro or in vivo.
- When interpreting nanoparticle interactions with biological cells and organisms, it is important to remember that living systems may appear normal and be capable of growth and function, but they may be genetically altered in subtle ways following nanoparticle exposure, which can produce serious consequences at some time in the distant future. Conversely, other cells that seem to be damaged may,

in time, recover from nanoparticle exposure and function normally in the absence of the nanoparticles.

- Seemingly static deposits of carbon-, gold-, or QD-nanoparticles within liver, lymph or other tissues/cells may be benign for some time, but can eventually degrade due to cellular processes or nanoparticle disaggregation, releasing intracellular toxic material.

Making predictions about nanoparticle biocompatibility that are based on the physicochemical attributes of that nanoparticle alone, or on limited experimentation (without looking at cell ultrastructure, cell viability and genetic changes), may result in conclusions that are not completely correct, but may be selectively true for a specific type of cell, nanoparticle or experimental design. This could present problems if the results were to be applied to other cells or to in vivo situations in various mammals. The answer to safe use and handling of nanoparticles lies in careful testing of nanomaterials under various conditions, using biological models for testing that represent the cells, tissues or organisms for the intended application, and using a multidisciplinary approach to examining cell responses so that it is possible to specifically know the fate of nanoparticles intracellularly, and the cell viability with any given nanoparticle.

The use of nanoparticles for creating better medical and imaging capabilities, producing new industrial and electronic components, and improving structural characteristics for many commercial products will provide new and better technological advances. Understanding their reactive properties and how living cells and organisms may interact with these new materials can eliminate future concerns about toxic responses to nanoparticles that could result during manufacture, disposal and utilization by the medical profession, manufacturers, researchers and the general public. The only weapon that we have to insure that these new materials are well designed and safely used is to question and test each new nanoparticle to make sure that it has been designed for safety (with maximum biocompatibility) during handling, use and disposal.

Acknowledgements This research and presentation could not have been done without the collaboration of my colleagues who helped to make our research in this area of biological–nanoparticle interactions possible and exciting. Special thanks go to Dr. John Warren of the Instrumentation Division, and Center for Functional Nanomaterials at Brookhaven National Laboratory who participated in all of the microscopy, nanoparticle imaging and microanalytical experiments, and to Drs. Oleg Gang, Mathew Maye, and Dmytro Nykypanchuk of the Center for Functional Nanomaterials (Brookhaven National Laboratory), and Daniel van der Lelie, Safiyh Taghavi, Betsy Sutherland and Paula Bennett of the Biology Department at Brookhaven National Laboratory. I would also like to thank Drs. Wynne Schiffer, Richard Ferrieri and the scientists and students (esp. Joseph Carrion) involved with the MicroPET research group at Brookhaven National Laboratory; and Drs. Mandakini Kanungo and Stanislaus Wong of Stony Brook University and Brookhaven National Laboratory, who oxidized some of our Carbolex nanotubes. This manuscript has been authored by Brookhaven Science Associates, LLC under Contract No. DE-AC02-98CH10886 with the US Department of Energy. The US Government retains, and the publisher, by accepting the paper for publication, acknowledges, a worldwide license to publish or reproduce the published form of this manuscript, or allow others to do so, for the US Government purposes.

References

1. M. Ross, E. Reith and L.Romwell, Histology a Text and Atlas, Williams & Wilkins. Baltimore (1989)
2. S. Moghimi, A. Hunter and J. Murray. Pharm. Rev. **53** (2), 283 (2001)
3. A. Nel et al.: Science **311**(5761), 622 (2006)
4. B. Panessa-Warren et al.: J. Phys: Condens. Matter. **18**, S2185 (2006), DOI: 10.1017/5143192760707065l/
5. B. Panessa-Warren et al.: Int. J. Nanotechnology **5**(1), 55 (2008)
6. A. Vertegel et al.: Langmuir **20**(26), 6800 (2004)
7. C. Salvador-Morales et al.: Mol. Immunol. **43**(3), 193 (2006)
8. P. Cherukuri et al.: PNAS **103**(50), 18882 (2006)
9. G. Oberdorster et al.: Part. Fibre Toxicol. **2**(8), 1 (2005)
10. D. Cui et al.: Toxicol. Lett. **155**(1), 73 (2005)
11. C. M. Sayes et al.: Biomaterials. **26**(36), 7587 (2005)
12. C. M. Sayes et al.: Toxicol. Lett. **161**(2), 135 (2006)
13. N. Monteiro-Riviere et al.: Toxicol. Lett. **155**, 377 (2005), Doi: 10.1016/jtoxlet.2004.11.004
14. B. Qualman and M. Kessels. Int. Rev. Cytol. **220**, 93 (2002)
15. G. Oberdorster, E. Oberdorster and J. Oberdorster. Environ. Health Perspect. **113**(7), 823 (2005)
16. A. Nel. Science **308**, 804 (2005)
17. A. Shvedova et al.: Am. J. Physiol. Lung Cell Mol. Physiol. **289**, 698 (2005)
18. G. Xiao et al.: J. Biol. Chem. **278**, 50781 (2003)
19. D. Lidke and D. Arndt-Jovin. Physiol. **19**(6), 322 (2004)
20. D. Lidke et al.: Nat. Biotechnol. **22**(2), 198 (2004)
21. Consumer Reports, July 2007:40–45; Woodrow Wilson International Scholars' Project on Emerging Nanotechnologies. Nanotechnology-based Consumer Products Inventory, http://www.nanotechproject.org/consumer/analysis.html, Oct.2007
22. S. Bellucci. Phys. Stat. Sol. C. **2**(1), 34 (2005)
23. M. Bronikowski et al.: J. Vac. Surf. Films **19**(4), 1800 (2001)
24. A. D. Maynard et al.: J. Nanopart. Res. **9**, 85 (2007)
25. T. Park et al.: J. Mater. Chem. **16**, 141 (2006)
26. A. Krause,W. Carley and W. Webb. J: Histochem. Anad Cytochem. **32**(10), 1084 (1984)
27. A. A. Shvedova et al.: J. Toxicol. Env. Health- Part A. **66**, 1909 (2003)
28. V. E. Kagan et al.: Toxicol. Lett. (2006) Doi:10.1016/j.toxlet.2006.02.001
29. D. B. Warheit et al.:Toxicol. Sci. **77**, 117 (2004)
30. C. W. Lam et al.: Toxicol. Sci.**77**, 126 (2004)
31. G. Oberdorster et al.: J. Toxicol. Environ. Health A. **65**, 1531 (2002)
32. P. Wick et al.: Toxicol. Lett. **168**(2), 121 (2007)
33. K. Pulskamp et al.: Toxicol. Lett. **168**(1), 58 (2007)
34. F. Tian et al.: Toxicol. In Vitro. **20**(7), 1202 (2006)
35. M. Becker et al.: Adv. Material. **19**, 939 (2007)
36. A. Porter et al.: Nat. Nanotechnol. **2**, 713 (2007)
37. B. Ballou et al.: Bioconjugate Chem. **15**(1), 79 (2004)
38. A. Hoshino et al.: Nano. Letters **4**(11), 2163 (2004)
39. X. Gao et al.: Curr. Opin. Biotechnol. **16**, 63 (2005)
40. J. Jaiswal et al.: Natur. Biotechnol. **21**, 47 (2003)
41. B. Dubertret et al.: Science **298**(5599), 1759 (2002)
42. A. M. Derfus, W. C. Chan and S. N. Bhatia. Nano Lett. **4**(1), 11 (2004)
43. I. Rikans and T. Yamano. J. Biochem. Mol. Toxicol. **14**, 110 (2000)
44. W. Schiffer, R. Ferrieri et al.: Brain 2007 and Brain Pet Proceedings 2007, Proc.8th Internat. Conf. on Quantification of Brain Function with PET, Osaka Japan, PSI-6M (2007)
45. P. Vaska et al.: Int. Rev. Neurobiol. **73**, 191 (2006)

46. B. Panessa-Warren et al.: Nanoscience & Nanotechnology Proceedings, International Symposium, 37 (2006)
47. J. Godt et al.: J. Occup. Med. Toxicol. **1**, 22 (2006)
48. H. Shimada, T. Funakoshi and M. Waalkes. Toxicol. Sci. **53**, 44 (2000)
49. R. Hardman. Environ. Health Perspectioves **114**(2) 165 (2006)
50. R. Ferrieri (personal communication) Brookhaven National Laboratory, Upton, NY
51. X. Sun et al.: Biomacromolecules. **6**, 2541 (2005)
52. S. Takeoka et al.: Biochem. Biophys. Res. Commun. **191**, 765 (2002)
53. The Merck Manual, 15th Ed., (R. Berkow, Ed) Merck Sharp & Dohme Research Laboratories, Merck & Co., Inc. Rahway, NJ. 1244, (1987) Rheumatoid Arthritis
54. J. Hillyer and R. Albrecht. Microsc. Microanal. **4**, 481 (1999)
55. M. Maye et al.: J. Am. Chem. Soc. **128**, 14020 (2006)
56. M. Maye et al.: Small (2007) DOI: 10.1002/smll.200700357
57. H. Xiong, D. van der Lelie and O. Gang. J. Am. Chem. Soc. (2008)
58. J. Li et al.: Adv. Material. **20**, 138 (2007)
59. M. Tsoli et al.: Small **1**, 841 (2005)
60. B. Chithrani, A. Ghazani and W. Chan. Nano Lett. **6**(4), 662 (2006)
61. N. Pernodet et al.: Small **2**, 766 (2006)
62. P. Cossart, P. Boquet, S. Normark and R. Rappuoli. Cell. Microbiol., ASM Press, Washington, D. C. 131 (2000)

Carbon Nanotubes Toxicity

Stefano Bellucci

Abstract We describe current and possible future developments in nanotechnology for biological and medical applications. Nanostructured, composite materials for drug delivery, biosensors, diagnostics and tumor therapy are reviewed as examples, placing special emphasis on silica composites. Carbon nanotubes are discussed as a primary example of emerging nanomaterials for many of the above-mentioned applications. Toxicity effects of this novel nanomaterial are discussed and the need for further study of potential hazards for human health, professionally exposed workers and the environment is motivated.

1 Introduction

The purpose of nanotechnology is not merely creating useful or functional materials and devices by manipulating matter at the nanometer length scale, but most importantly exploiting novel properties of materials which arise just owing to the nanoscale. Simply meeting the length scale criterion of 1–100 nm is not really nanotechnology, rather it is a necessary condition; the corresponding sufficient condition consists in taking advantage of novel (physical, chemical, mechanical, electrical, optical, magnetic, etc.) properties that result solely because of going from bulk to the nanoscale.

Since 2001, when the U.S. announced a National Nanotechnology Initiative (NNI) aimed at creating a dedicated program to explore nanotechnology (see http://www.nano.gov/), many other countries, including the EU (see e.g. http://www.euronanoforum2007.eu/), Japan (see www.nanonet.go.jp), China (see e.g. http://www.sipac.gov.cn), followed up with their own nanotechnology research programs. It is important to recall the special character of nanotechnology as a both pervasive and enabling technology, with potential impact in all sectors of the economy: Electronics, computing, data storage, materials and manufacturing, health and medicine,

Stefano Bellucci
INFN-Laboratori Nazionali di Frascati, Via E. Fermi 40, 00044 Frascati, Italy,
e-mail: bellucci@lnf.infn.it

S. Bellucci (ed.), *Nanoparticles and Nanodevices in Biological Applications*. Lecture Notes in Nanoscale Science and Technology 7,

energy, transportation, environment, national security, space exploration and others. In this article, we focus on nanotechnology applications in the biomedical sector.

Carbon nanotubes (CNTs) are an example of a carbon-based nanomaterial [1], which has won enormous popularity in nanotechnology for its unique properties and applications [2]. CNTs have physicochemical properties that are highly desirable for use within the commercial, environmental, and medical sectors. With the inclusion of CNTs to improve the quality and performance of many widely used products, as well as potentially in medicine, it is likely that occupational and public exposure to CNT-based nanomaterials will increase dramatically in the near future. Hence, it is of the utmost importance to explore the yet almost unknown issue of the toxicity of this new material. Here, we compare the toxicity of pristine and oxidized multi-walled carbon nanotubes on human T cells and find that the latter are more toxic and induce massive loss of cell viability through programmed cell death at doses of 400 μg/ml, which corresponds to approximately 10 million carbon nanotubes per cell. Pristine, hydrophobic, carbon nanotubes were less toxic and a ten-fold lower concentration of either carbon nanotube type was not nearly as toxic. Our results suggest that carbon nanotubes indeed can be very toxic at sufficiently high concentrations and that careful toxicity studies need to be undertaken particularly in conjunction with nanomedical applications of carbon nanotubes.

2 Nanotechnology for Tumor Therapy

Over the past three and a half decades, since the beginning of the U.S. National Cancer Initiative in 1971, there have been major advances in the diagnosis and treatment of cancer. However, the severe toll cancer continues to impose on our society represents one of the major healthcare concerns of our nations. One out of every two men and one out of every three women in their lifetime will be confronted with a cancer diagnosis. Conventional treatments currently rely heavily upon radiation and chemotherapy, which are extremely invasive and painstakingly plagued by very serious side effects. Nanotechnology yields the hope for new methods for a noninvasive therapy, capable of minimizing side effects. One of the promising approaches consists in the targeted destruction of cancerous cells using localized heating.

The use of thermal cancer therapies is beneficial in many respects over the conventional tumor removal by surgery. Indeed, normally, most thermal approaches have a very small degree of invasiveness; they are relatively simple to perform and may enable physicians to treat tumors embedded in vital regions where surgical removal is unfeasible. Ideally, the activating energy to heat the tumor would be targeted on the embedded tumor with minimal effect on surrounding healthy tissue. Unfortunately, conventional heating techniques such as focused ultrasound, microwaves, and laser light do not discriminate between tumors and surrounding healthy tissues. Thus, success has been modest, and typically, treatments result in some damage to surrounding tissue.

Recent work suggests that nanostructures designed to attach to cancerous cells may provide a powerful tool for producing highly localized energy absorption at the sites of cancerous cells. Indeed, work since 2003 at La Charité Hospital in Berlin [3], with scientists at the F. Schiller-Universität Jena, showed that magnetic nanoparticles interstitially injected directly into the tumor, and heated with radio-frequency radiation [4], can destroy cancer cells in a human brain tumor and are also believed to enhance the effects of subsequent radiation therapy. Nanoparticles localize on the tumor due to a special biomolecularly modified outer layer – leaving the surrounding healthy tissue with minimum damage.

In this way, it was proven that iron oxide nanoparticles, with diameters 10,000 times smaller than that of a human hair, can be introduced inside cancer cells and then treated in such a way as to produce a significant damage to tumor cells in order to fight a particularly aggressive form of brain cancer called glioblastoma, although the method can be employed to treat other forms of the disease. The procedure involves coating the iron oxide nanoparticles with an organic substance, such as the sugar glucose, before injecting them into the tumor. Cancer cells, having a fast metabolism and correspondingly high energy needs, are much more eager to eat up the sugar-coated nanoparticles, in comparison with healthy cells, which appear minimally or not at all affected (Fig. 1).

In this selective procedure, the magnetic field is responsible for the heating up of the nanoparticles in the cancerous tissue, reaching temperatures up to 45°C (Fig. 2), with the aim of destroying many of the tumor cells or at least to weaken them to such an extent that conventional methods, e.g. radiation or chemotherapy, can more easily and effectively get rid of them.

The treatment, known as magnetic fluid hyperthermia, was successfully used to prolong the life of laboratory rats which were implanted with malignant brain tumors (Figs. 3 and 4). Rats receiving nanotherapy lived four times as long as rats receiving no treatment.

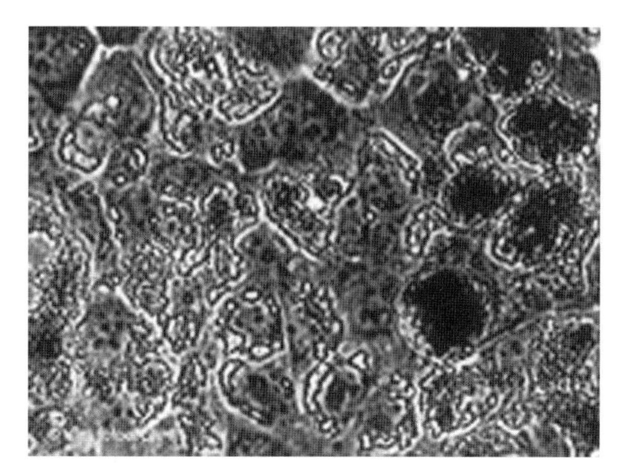

Fig. 1 The image shows nanoparticles surrounding cancer cells. Image source: MFH Hyperthermiesysteme GmbH and MagForce Applications GmbH, Berlin, Germany

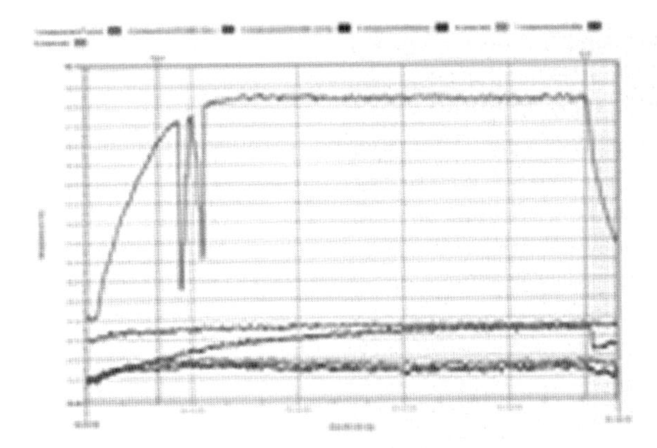

Fig. 2 The treatment is automatically recorded with the temperature of the tumor (*top curve in the diagram*) and other body-temperatures registered. Image source: MFH Hyperthermiesysteme GmbH and MagForce Applications GmbH, Berlin, Germany

Then, the therapy was given to 15 patients suffering from *Glioblastoma multiforme*, the most common primary brain tumor and the most aggressive form of brain cancer (with a 6–12 months life expectancy prognosis in humans).

The treatment is particularly attractive to doctors working with tumors in the brain since the nanoparticles can be targeted on the cancerous tissue, so that the therapy turns out to be ideal for curing tumors that lie outside the reach of conventional surgical treatment, such as those situated deep in the brain or in regions that are responsible for essential tasks like speech or motor functions.

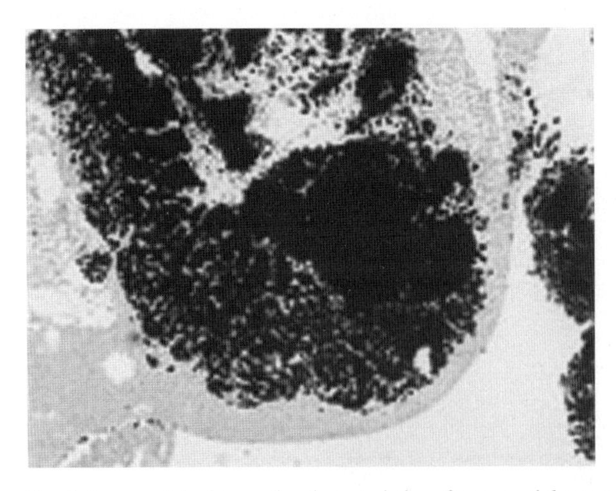

Fig. 3 In pre-clinical tests, the characteristics of nanoparticles were optimized; shown: accumulation of nanoparticles in tumor tissue (RG-2 glioblastoma of the rat). Image source: MFH Hyperthermiesysteme GmbH and MagForce Applications GmbH, Berlin, Germany

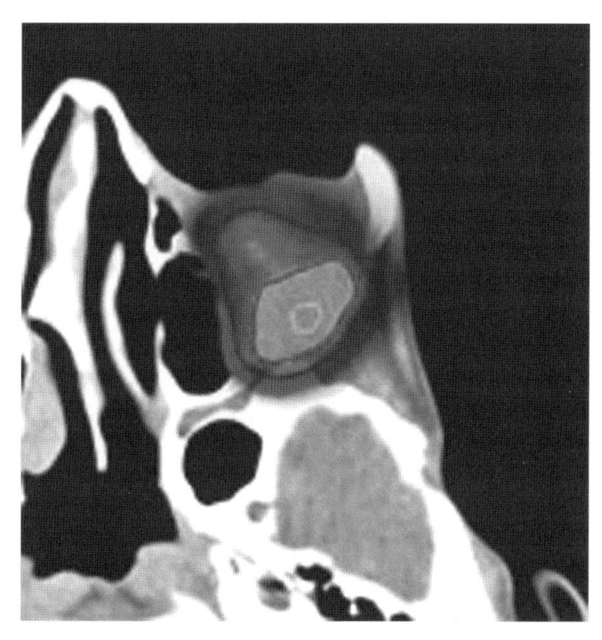

Fig. 4 A precise thermotherapy of target areas in almost every body region is possible (here, thermotherapy of the orbita up to a maximum temperature of 49°C). Image source: MFH Hyperthermiesysteme GmbH and MagForce Applications GmbH, Berlin, Germany

In principle, the hyperthermia therapy is not limited to just various types of brain cancer. Since breast tumors do not lie in the immediate vicinity of essential organs, one can hope to apply the treatment, heating the cancerous tissue up to yet higher temperatures, in order to get a very effective cure of breast tumor, which may even be combined with parallel treatments relying upon conventional radiation therapy and chemotherapy.

However, one should bear in mind a caveat: keeping the amount of metal injected into the body stay under a certain level is the way to maintain the danger of "nanopoisoning" at a relatively low level. N.B.: After all, it should be remembered that nanoparticles are already used routinely in magnetic resonance therapy for the diagnosis of liver tumors.

After the therapy, nanoparticles do not have to be removed and are slowly metabolized. Since, so far, no harmful side effects from thermotherapy with magnetic nanoparticles could be observed, neither on animals nor on human beings, in 2004 nanoparticles have started to be applied for treating human prostate carcinomas at the Clinic for Urology, Charité – University Medicine, Berlin, Germany.

Moving to countries outside Europe, in order to complete the survey, we observe that in Japan [5], work at Nagoya University with magnetite cationic liposomes (MCLs) combined with heat shock proteins has shown great potential in cancer treatment as well. Using MCLs, one locally generates heat in a tumor by placing test mice in an alternating magnetic field and not cause the body temperature of the test animal to rise. After injection of MCLs and application of a magnetic field, tumor and body temperature differed by 6°C. The combined treatment strongly inhibited tumor growth over a 30-day period and complete regression of tumors was observed in 20% of the mice.

Finally, in the U.S., researchers at Rice University recently reported work on mice in which gold-coated nanoparticles treated to attach to cancerous cells were heated using infrared radiation. Sources of infrared radiation can be tuned to transmit at a narrow band of electromagnetic frequencies. Additionally, the "nanoshells" size can be changed to absorb a particular infrared radiation frequency. Hence, one can choose a frequency of the infrared radiation that couples with the gold-coated nanoparticles, while at the same time, does not couple with the tissue of the body, thus enabling the selective destruction of cancerous cells and tumors [6]. The results of a preliminary experiment with mice treated with the nanoshells-infrared radiation therapy have proven to be very encouraging.

In conclusion, we can say that progress in nanotherapy, obtained by several groups worldwide, using independent techniques, shows the global interest in nanoscience and its potential application to innovative medical technologies. This justifies the expectation that nanotechnology will soon yield a powerful tool for treating cancer.

3 Nanotechnology for Diagnostics and Drug Delivery (The source for the material contained in this section is: Strem Chemicals)

Magnetic nanoparticles have been used as markers in biomedical diagnostics. In fact, due to the fact that bound and unbound nanoparticles have different magnetic relaxation times, biochemical binding reactions can be detected by means of a SQUID-high resolution measurement technique (Fig. 5). Magnetic relaxation immunoassays were realized by means of this technique. Also, in vivo-applications of magneto-relaxometry seem possible, e.g. in cancer diagnostics.

However, the use of magnetic nanoparticles is not limited to the abovementioned applications, but can also be extended for achieving drug or radiation delivery. In particular, small magnetic particles can be engineered to carry therapeutic chemicals

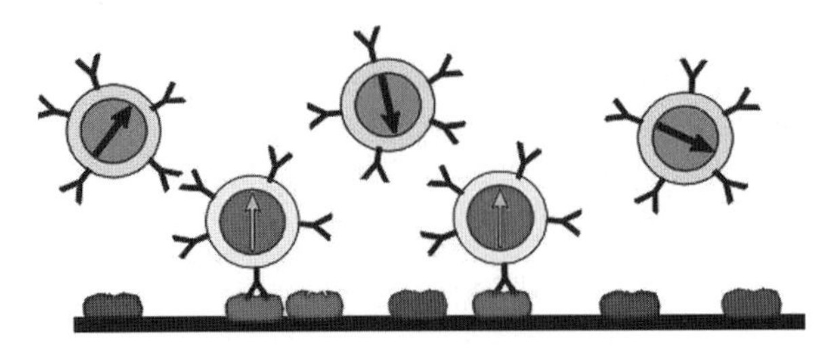

Fig. 5 Immobilization of magnetic nanoparticles by antibody-antigen coupling (Contact: Dietmar.Eberbeck@ptb.d)

or radiation for tumor control. Because they are magnetic, the particles can be guided by an external magnetic field and can be forced to move with or against the flow of blood in veins or arteries or held in a fixed position once they have been conveyed to a target organ, and possibly retrieved when treatment has ended. Work is in progress at Argonne Nat. Lab. and the University of Chicago.

One is, of course, interested in using biosensors and biolabels to understand living cells. Nanotechnology has the potential to increase our ability to understand the fundamental working of living cells. Many potential applications for nanomaterials as biosensors and biolabels are under investigation. They have found use in cellular studies, enhanced spectroscopic techniques, biochips, and protein and enzyme analysis. Fluorescent nanoparticles can be used for cell labeling and magnetic nanoparticles may be utilized as sensors. Multi-color labeling of both fixed and living cells with fluorescent nanoparticles conjugated with biological ligands that specifically bind against certain cellular targets enables the recording of diffusion pathways in receptor cells.

The uptake of nanoparticles into the vesicular compartments around the nucleus of cells can be used to label the cells so that their pathway and fate can be followed. The nanoparticles exhibit reduced photobleaching as compared to traditional dyes and are passed on to daughter cells during cell division, therefore allowing for much longer term observation. Magnetic nanoparticles can also act as sensors for assessing how external stresses affect changes in intracellular biochemistry and gene expression.

We can ask ourselves, how can nanotechnology improve medical diagnostics? Naturally, the early detection of a disease remains the primary goal of the medical community. Nanotechnology holds great promises for enabling the achievement of this goal. Nanoparticles, in particular, have exhibited tremendous potential for detecting fragments of viruses, pre-cancerous cells, disease markers, and indicators of radiation damage. Biomolecule coated, ultra small, superparamagnetic iron oxide (USPIO) particles injected in the blood stream recognize target molecular markers present inside cells and induce a specific signal for detection by magnetic resonance imaging (MRI). This technology may allow for detection of individual cancer cells months or years earlier than traditional diagnostic tools, which require the presence of hundreds of cancer cells.

In this respect, we can also address the issue of how biobarcode amplification assays (BCA) can use nanoparticles in disease detection. A nanoparticle-based BCA utilizes gold nanoparticles and magnetic microparticles attached to large numbers of DNA strands and antibodies for a specific disease marker. The marker binds to the nano- and microparticles forming a complex that is separated from the sample using a magnetic field. Heating the complexes releases the DNA barcodes, which emit an amplified signal due to their large numbers. This BCA technology has been applied to the detection of markers for Alzheimer's disease and is being investigated for numerous others.

There are also stimulating suggestions that nanotechnology can improve targeted drug delivery. Targeted drug delivery systems can convey drugs more effectively and/or more conveniently, increase patient compliance, extend the product life cycle, provide product differentiation, and reduce health care costs. Drug delivery systems that rely on nanomaterials also allow for targeted delivery of compounds characterized by low oral bioavailability due to poor water solubility, permeability and/or instability and provide for longer sustained and controlled release profiles. These technologies can increase the potency of traditional small molecule drugs in addition to potentially providing a mechanism for treating previously incurable diseases.

There are many other applications for nanomaterials in the medical and pharmaceutical sector, which we only have the possibility, owing to limitations in the length of these notes, to merely list. Areas currently under investigation include gene therapy, antibacterial/antimicrobial agents for burn and wound dressings, repair of damaged retinas, artificial tissues, prosthetics, enhancing signals for magnetic resonance imaging examinations, and as radio frequency controlled switching of complex biochemical processes.

4 Carbon Nanotubes

The development of nanomaterials is currently underway in laboratories worldwide for medical and biotechnological applications including gene delivery [7, 8] drug delivery [9, 10] enzyme immobilization [11, 12] and biosensing. [13, 14]. The most commonly used materials are gold [15], silica and semiconductors. Silica nanoparticles have been widely used for biosensing and catalytic applications due to their large surface area-to-volume ratio, straightforward manufacture, and the compatibility of silica chemistry with covalent coupling of biomolecules [16–18].

A key challenge in nanotechnology is the more precise control of nanoparticle assembly for the engineering of particles with the desired physical and chemical properties. Much research is currently focused on CNT as a promising material for the assembly of nanodevices, based upon new CNT–composite materials, such as CNT with a thin surface cover [19, 20] or CNT bound to nanoparticles [21–24], in order to tailor their properties for specific applications.

In this section, reviewing the results reported in [25], we present the tunable synthesis of multi-walled CNT–silica nanoparticle composite materials. Instead of coupling prefabricated silica nanobeads to CNT, we chose to grow the silica nanobeads directly onto functionalized multi-walled CNT by reaction of tetraethyl- or tetramethyl-orthosilicate (TEOS or TMOS) with a functionalized CNT precursor, prepared by coupling aminopropyltriethoxysilane (APTEOS) to a functionalized multi-walled CNT through a carboxamide bond, using a water-in-oil microemulsion to strictly control the nanobead size.

The body of the ideal multi-walled CNT is formed by several nested and straight cylindrical graphene sheets. In reality, nanotubes usually appear curved and have topological defects. Under strong oxidizing conditions (conc. HNO_3), nanotubes can be cut into shorter and straighter pieces having carboxylic acid groups at both their tips and at imperfections on their walls [26]. We oxidized multi-walled CNT with outer diameters of ca. 20–40 nm and lengths of 5–10 mm (NanoLab, Inc.,

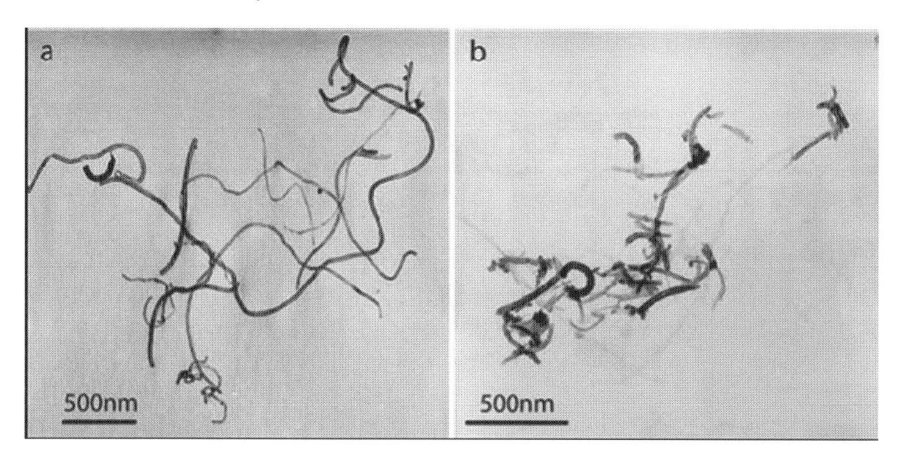

Fig. 6 Multi-walled carbon nanotubes before (**a**) and after (**b**) oxidation in nitric acid

Newton, MA, USA) (Fig. 6a) by refluxing in concentrated HNO_3 for 6 hours, followed by several washes with distilled water. The oxidized CNT (CNT-COOH) were shorter and straighter (Fig. 6b).

Their carboxylic acid groups greatly facilitated their dispersion in aqueous solutions, as well as their further functionalization (Fig. 7a). A detailed description of the procedure for generating the activated CNT precursor (CNT-APTEOS) to the composite from CNT-COOH by activation of its carboxylic acid groups can be found in [25].

Using the procedure described in [25], we obtained new CNT–nanocomposites consisting of CNT with covalently attached silica nanobeads (Fig. 8). Non-oxidized CNT (with negligible COOH content) did not support any composite formation (not shown). The inverse microemulsion system resulted in nanobeads covalently linked to the CNT only at locations functionalized with triethoxy-silane groups, while the bare graphitic wall of the pristine CNT did not associate with reverse micelles. Transmission electron microscopic (TEM) images revealed morphologies indicative of different nanobead diameters. Small nanoparticles were found to decorate the walls and ends of the CNT prepared using TMOS as precursor (Fig. 8a–c). In many cases, small nanoparticle aggregates were observed to be associated with the CNT (Fig. 8c), as expected for the high density of functional groups on the CNT. Under the conditions used for synthesis of larger nanoparticles, CNT were either decorated by individual nanobeads (Fig. 8d–f) or had a uniform silica coating around the entire CNT (Fig. 8g and h). We also observed some functionalized CNT that appeared to have silica within their tubes (Fig. 8i). The internal presence of silica was not observed with the non-treated nanotubes. Further work is in progress to better understand the filling mechanism.

In summary, we covalently coated carbon nanotubes with silica nanoparticles of different sizes. Perhaps, the most valuable feature of our work [25] is that the architecture of the obtained assemblies can be largely controlled by varying the conditions in the synthesis. Thus, the length of CNT is regulated by the oxidation time

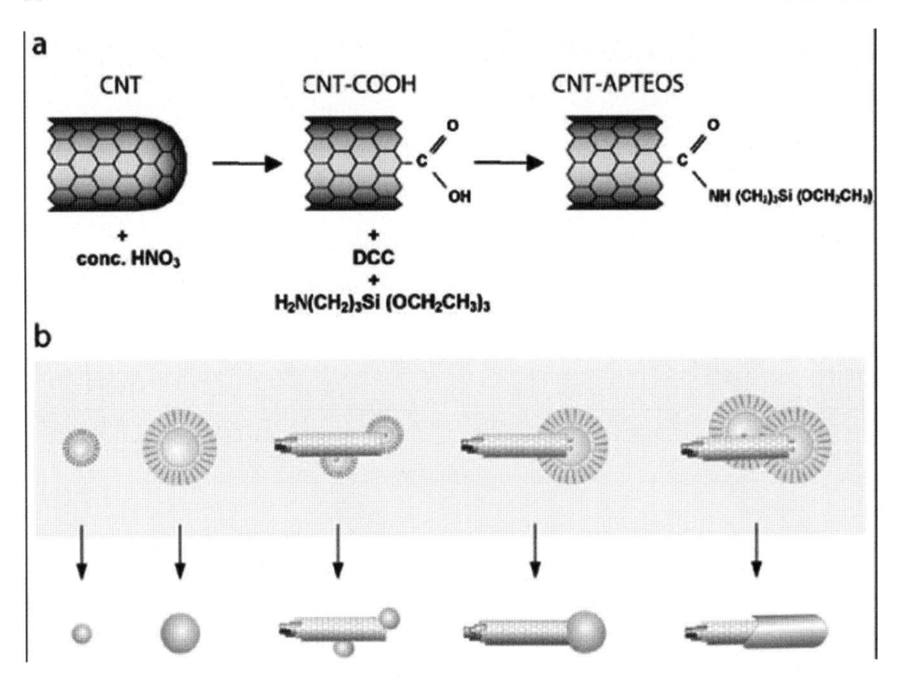

Fig. 7 Scheme for preparing the CNT–nanoparticle composite. (**a**) Oxidation and preparation of the CNT-APTEOS precursor. (**b**) Formation of silica nanobeads in reverse micelles in a water-in-oil microemulsion. Inclusion of CNT-APTEOS nucleates the formation of nanobeads on the covalently linked propyltriethoxysilane groups (dots inside the micelles) by reaction with TEOS or TMOS

(Fig. 6) and the size of the nanobeads by using microemulsion conditions that yield micelles of a particular size. Indeed, silica nanobeads were prepared in a water-in-oil microemulsion system in which the water droplets served as nanoreactors [27, 28]. The size of the final nanospheres was mainly regulated by the dimension of the water droplets, and therefore, by the molar ratio of water to surfactant (w). Smaller nanobeads were prepared by reducing w. Furthermore, the dimension of the final product can be controlled by varying the molar ratio of water to precursor (h), the molar ratio of precursor to catalyst (n), by choosing the reactivity of the precursor, and the reaction time and temperature. The values of the variable parameters (w,h,n) used can be found in Table 1 of [25].

Because the chemical properties of the silica surface are particularly versatile and silica can be doped with fluorescent [29], magnetic [30] or biological macromolecules [31], nanostructures with a wide range of morphologies suitable for different applications can be obtained. We anticipate that further refinement of our water-in-oil microemulsion approach for creating novel nanostructures combined with procedures for isolating discrete products will allow us to combine different nanostructures into higher order assemblies that could be useful for a variety of applications, including providing an interface between living cells and biosensor arrays.

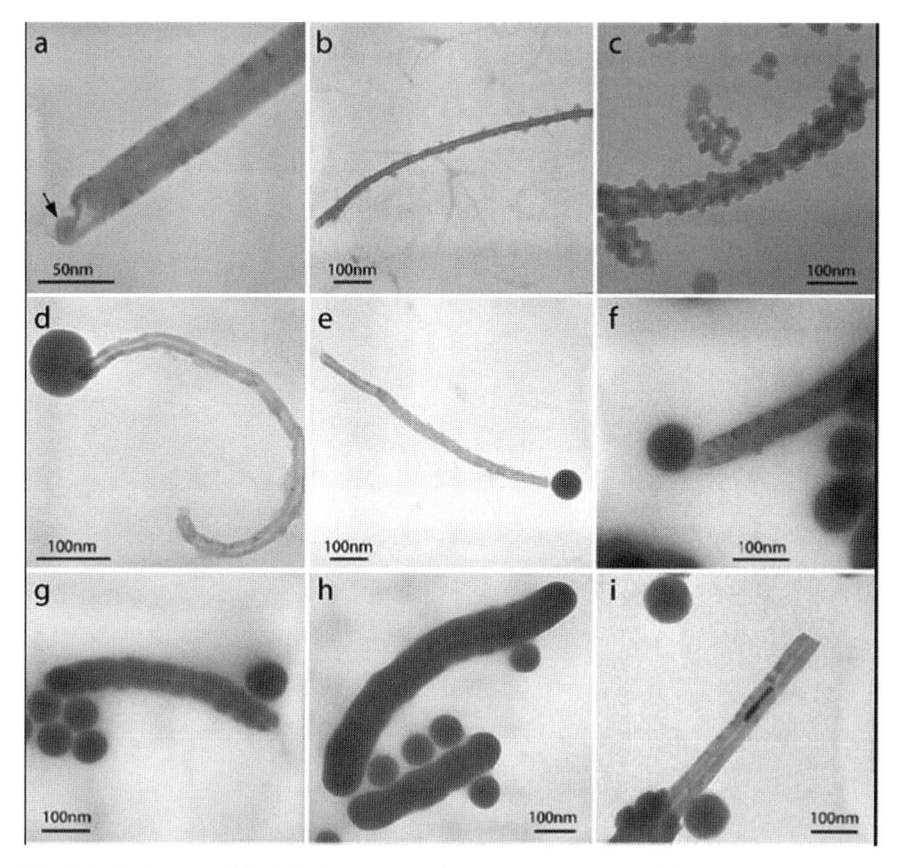

Fig. 8 TEM images of the CNT–nanocomposites prepared using conditions for small (**a–c**) or large (**d–i**) silica nanobeads. The *arrow* in panel (**a**) indicates a nanobead at the tip of the CNT. The *arrow* in panel (**i**) indicates a polymerized silica inside a CNT

5 Supramolecular Nanostructures

In [32], we constructed and characterized supramolecular nanostructures consisting of ruthenium-complex luminophores, which were directly grafted onto short oxidized single-walled carbon nanotubes or physically entrapped in silica nanobeads, which had been covalently linked to short oxidized single-walled carbon nanotubes or hydrophobically adsorbed onto full-length multi-walled carbon nanotubes. These structures were evaluated as potential electron-acceptor complexes for use in the fabrication of photovoltaic devices, and for their properties as fluorescent nanocomposites for use in biosensors or nanoelectronics.

The carboxylic acid groups of oxidized SWCNT which originated from the nitric acid-oxidation, were covalently tethered to the ruthenium-complexes (Fig. 9A) or luminophore-doped silica nanobeads (Fig. 9B), whereas the full-length MWCNT had the ruthenium complex-doped silica nanobeads introduced onto their surfaces

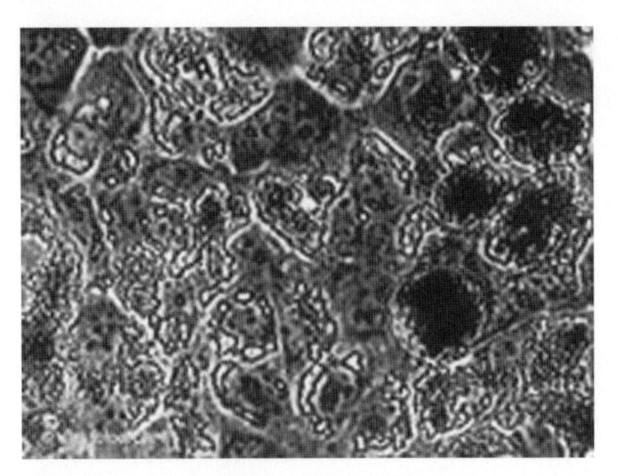

Fig. 9 Scheme for preparing the supramolecular nanostructures. **A**: Ruthenium-complex luminophores were directly grafted onto oxidized SWCNT to form a supramolecular donor–acceptor nanostructure. **B**: Ruthenium-complex luminophore-doped silica nanobeads were covalently linked to short oxidized SWCNT to form a supramolecular fluorescent nanostructure. **C**: Ruthenium-complex luminophore-doped silica nanobeads were hydrophobically adsorbed onto full-length MWCNT via π-π interactions to form a supramolecular fluorescent nanostructure with CNT having the π-electronic structure intact

by hydrophobic adsorption via π–π interactions to maintain the intact CNT π-electronic structure.

The absorbance spectrum of $Ru(ap)(bpy)_2$ in DMF exhibited a narrow peak at approximately 289 nm, which we attribute to a ligand-to-ligand $\pi \rightarrow \pi^*$ transition, and by two broad bands at approximately 375 and 460 nm, which we attribute to $d_{Ru} \rightarrow \pi^*$ ligand singlet metal-to-ligand charge-transfer transitions (Fig. 10A). On excitation at 460 nm, the steady state emission of $Ru(ap)(bpy)_2$ in DMF revealed an emission band centered at 606-nm (Fig. 10B).

The absorbance spectrum of $Ru(ap)(bpy)_2$-decorated SWCNT dissolved in DMF was broad and slightly blue-shifted compared to that from $Ru(ap)(bpy)_2$ in DMF to confirm that luminophores decorated the nanotube surface (Fig. 10A). The spectroscopic contribution of $Ru(ap)(bpy)_2$ grafted onto SWCNT was calculated by subtracting from the absorbance spectrum of the nanostructure that of oxidized SWCNT (dissolved in DMF), with matching absorption at 900 nm, because the spectroscopic contribution of the luminophore was absent for that wavelength. The peaks in the calculated absorbance spectrum of $Ru(ap)(bpy)_2$ grafted on SWCNT were centered at the same wavelengths and slightly broader compared to those of $Ru(ap)(bpy)_2$ in DMF. On excitation at 460 nm, the steady state emission spectrum of $Ru(ap)(bpy)_2$-decorated SWCNT in DMF revealed a strongly quenched ($> 98\%$) 606-nm photoluminescent peak compared to that of $Ru(ap)(bpy)_2$ in DMF (Fig. 10B). Emission spectra were collected after having matched the absorptions, at the 460-nm excitation wavelength, of free $Ru(ap)(bpy)_2$ and $Ru(ap)(bpy)_2$ grafted onto SWCNT. Excitation at 375 nm showed similar quenching to further confirm the formation of the linked supramolecular electron donor-acceptor complexes between

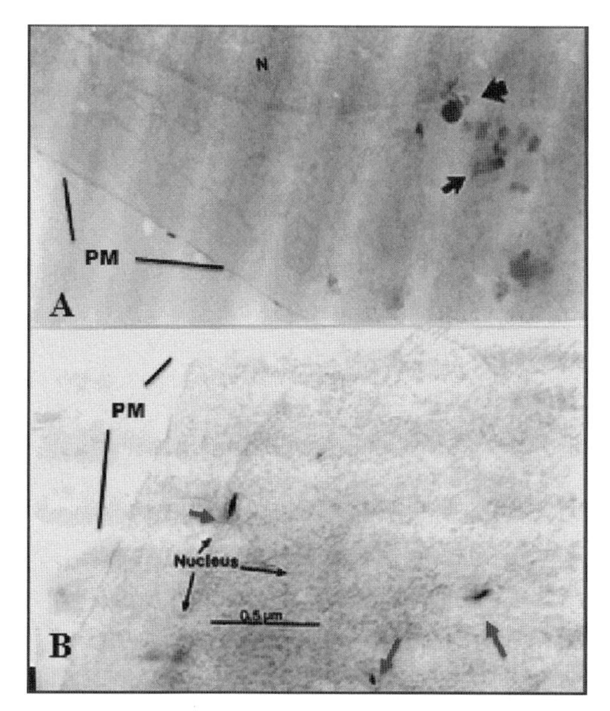

Fig. 10 Absorbance (**A**) and emission (**B**) spectra of Ru(ap)(bpy)$_2$ (*dotted line*) and Ru(ap)(bpy)$_2$-decorated SWCNT (*solid line*) dispersions in DMF. For the emission spectra, we matched the absorptions, at the 460-nm excitation wavelength, of free Ru(ap)(bpy)$_2$ and Ru(ap)(bpy)$_2$ grafted onto SWCNT, calculated by subtracting from the absorbance spectrum of the nanostructure that of oxidized SWCNT

the metalloorganic luminophores, which acted as electron-transfer agents, and the carbon nanotubes, which acted as electron acceptors. Addition of free luminophore to the nanocomposite dispersions increased their emission intensities. These increases in emission suggest that, at the low concentrations used, dynamic (collisional) quenching was not responsible for the observed fluorescence quenching, which could only be caused by photo-induced charge injection from the metal-to-ligand charge-transfer (both singlet and triplet) excited states of the luminophore into the conduction band of the SWCNT.

Transmission electron microscope images of fSNB showed uniform diameter (13 ± 1 nm) silica nanobeads. Absorbance spectra of both free and fSNB-encapsulated Ru(bpy) in EtOH were characterized by a narrow peak at approximately 290 nm and by a broad plateau at approximately 450 nm, which we attribute to a ligand-to-ligand $\pi \rightarrow \pi^*$ transition and a $d_{Ru} \rightarrow \pi^*$ ligand singlet metal-to-ligand charge-transfer transition, respectively. The peaks in the spectrum of the fSNB in EtOH were slightly red-shifted and broader compared to those of free Ru(bpy) (Fig. 11A). In steady state experiments, excitation of fSNB in EtOH at 452 nm produced an emission band that was enhanced, red-shifted and broader compared to that

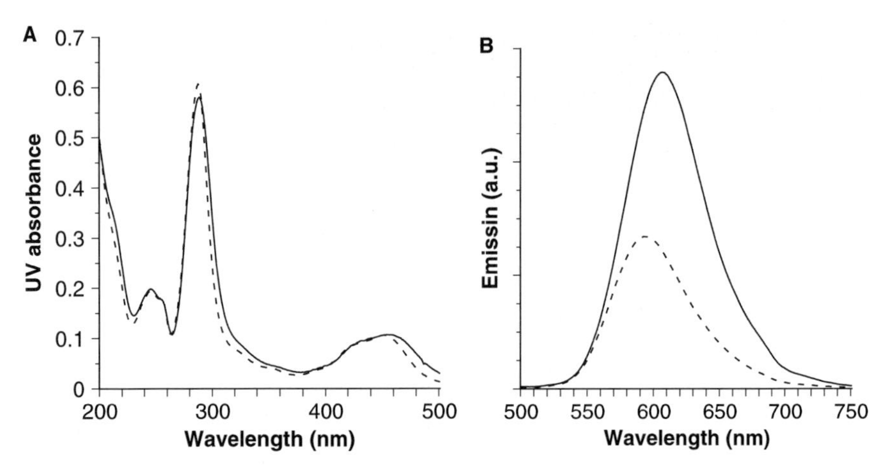

Fig. 11 Absorbance (**A**) and emission (**B**) spectra of Ru(bpy) (*dotted line*) and fSNB (*solid line*) dispersions in EtOH. For the emission spectra, we matched the absorptions, at the 452-nm excitation wavelength, of free Ru(bpy) and fSNB

of Ru(bpy) (Fig. 11B). The emission spectra were collected after having matched the absorptions, at the 452-nm excitation wavelength, of free Ru(bpy) and fSNB. These results suggest that the silica bead network may have affected the electrostatic environment surrounding the entrapped luminophores and that encapsulation protected Ru(bpy) from any dynamic self-quenching caused by collisional encounters.

The absorbance spectra of both fSNB-decorated SWCNT (Fig. 12A) and MWCNT (Fig. 12B) dissolved in EtOH showed broad peaks at approximately the same wavelengths as those from fSNB to confirm that fluorescent nanoparticles decorated the nanotube surfaces (Fig. 12C). The spectroscopic contribution of fSNB grafted onto the CNT surface was calculated by subtracting from the absorbance spectrum of the nanocomposite that of silylated CNT with matching absorption at 900 nm, as the spectroscopic contribution of free fSNB was absent for that wavelength. The peaks in the calculated absorbance spectra of fSNB grafted onto both (both SW and MW) CNT were slightly red-shifted compared to those of free fSNB in EtOH. On excitation at 452 nm, the steady-state emission spectrum of fSNB-decorated CNT exhibited slightly quenched ($< 5\%$), narrower and blue-shifted emission peaks compared to that of free fSNB (Fig. 12D). The emission spectra were collected after having matched the absorptions, at the 452-nm excitation wavelength, of free fSNB and fSNB grafted onto CNT.

The observed slight quenching could be addressed to the error in the calculation of fSNB grafted onto CNT and/or to the electron transfer from the luminophores close to the fSNB silica surface to the CNT. Therefore, the silica host was able to avoid the quenching of the fluorescence due to the charge injection from the metal-to-ligand charge-transfer excited states of the luminophore into the conduction band of the quencher (CNT), leading to the realization of a supramolecular fluorescent nanostructure useful for a large variety of applications ranging from biosensors to

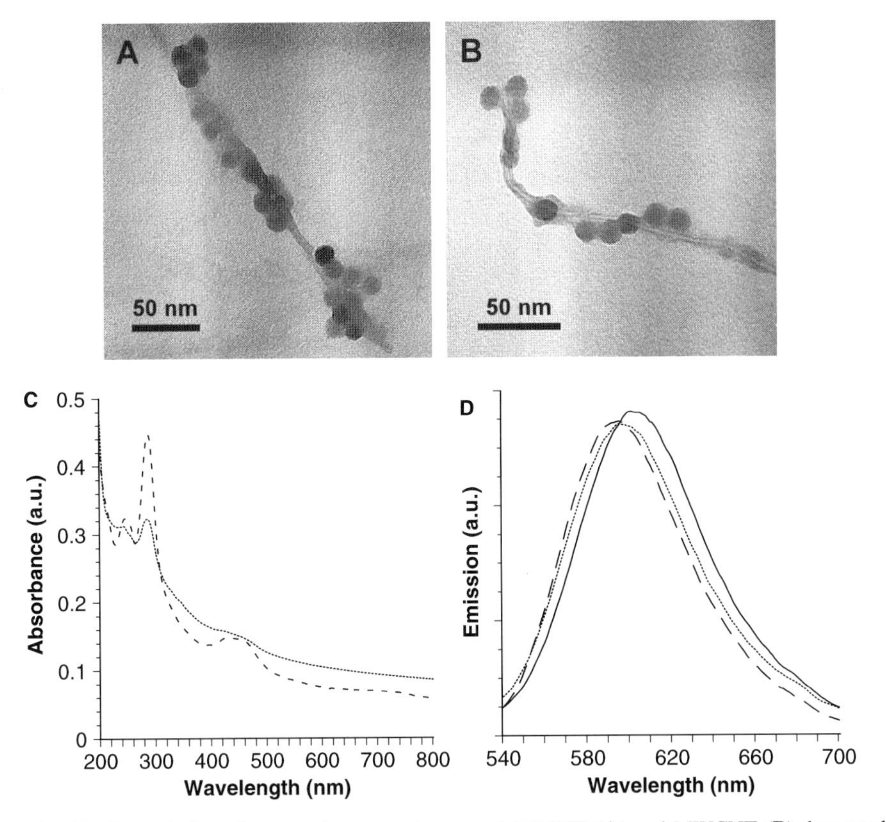

Fig. 12 Transmission electron microscope images of SWCNT (**A**) and MWCNT (**B**) decorated with fluorescent silica nanobeads. C: Absorbance spectra of fSNB-decorated SWCNT (*dashed line*) and fSNB-decorated MWCNT (*dotted line*) dispersions in EtOH. D: Emission spectra of fSNB (*solid line*), fSNB-decorated SWCNT (*dashed line*) and fSNB-decorated MWCNT (*dotted line*) dispersions in EtOH. For the emission spectra, we matched the absorption values at the 452-nm excitation wavelength, of free fSNB and fSNB grafted onto CNT, calculated by subtracting from the absorbance spectrum of the nanocomposite that of silylated CNT

electronics, especially in case of use of pristine full-length CNT that are characterized by intact π-electronic structure.

In summary, we synthesized in [32] three supramolecular nanostructures based on CNT and ruthenium-complex luminophores. The first nanostructure consisted of short oxidized SWCNT covalently decorated by ruthenium-complexes that act as light-harvesting antennae by donating their excited-state electrons to the SWCNT. This nanocomposite represents an excellent donor-acceptor complex, which may be particularly useful for the construction of photovoltaic devices based on metalloorganic luminophores. The second and the third nanostructures consisted of metalloorganic luminophore-doped silica nanobeads covalently linked to short oxidized SWCNT or hydrophobically adsorbed onto full-length MWCNT. In these nanocomposites, the silica network prevented the fluorescence quenching because excited-state electrons could not be readily donated to the CNT conduction band. Because the physical and chemical properties of the silica nanobeads are so versatile, and the π-electronic structure of the CNT can be kept intact by using a non-destructive modification of the nanotube structure, we consider these nanocomposites to have a promise for a variety of applications ranging from the biosensors to electronics.

6 Cellular Toxicity of Carbon Nanotubes

Very little is yet known about the toxicity of CNTs, which exist in many different forms and can be chemically modified and/or functionalized with biomolecules. Pristine single-walled CNTs are extremely hydrophobic tubes of hexagonic carbon (graphene) with diameters as small as 0.4 nm and lengths up to micrometers. Multi-walled CNTs consist of several concentric graphene tubes and diameters of up to 100 nm. These pristine CNTs are chemically inert and insoluble in aqueous media and therefore of little use in biological or medical applications. Due to the hydrophobicity and tendency to aggregate, they are harmful to living cells in culture [33, 34].

As we have seen in one of the above sections, for many applications, CNTs are oxidized in strong acid to create hydroxyl and carboxyl groups [35], particularly at their ends, to which biomolecules or other nanomaterials can be coupled [25]. These oxidized CNTs are much more readily dispersed in aqueous solutions and have been coupled to oligonucleotides, proteins, or peptides. Indeed, CNTs have been used as vehicles to deliver macromolecules that are not able pass through the cellular membrane by themselves into cells [36, 37].

Since little is yet known about the toxicity of CNTs, particularly of oxidized CNTs, we compared in [38] these two types of CNTs in a number of functional assays with human T lymphocytes, which would be among the first exposed cell types upon intravenous administration of CNTs in therapeutic and diagnostic nanodevices.

We found that, especially for high concentration (>1ng/ cell), CB are less toxic than pristine CNTs, therefore suggesting the relevance of the structure and topology (carbon black is amorphous) on the evaluation of the toxicity of a carbonaceous nanomaterial. Moreover, we found that oxidized CNTs are more toxic than pristine CNTs for both the analyzed concentrations, although they are considered better suited for biological applications. This may well be because they are better dispersed in aqueous solution and therefore reach a higher concentration of free CNTs at similar weight per volume values. We calculated that the less toxic amount of 40 μg/ml of CNTs is equal to an order of magnitude of 10^6 individual CNTs per cell in our experiments, based on an average length of 1 μm and a diameter of 40 nm, giving an average molecular mass of 5×10^9 Da.

While our results in [38] do not imply that CNTs should be abandoned for biological or medical purposes, our study sets an upper limit for the concentrations of CNTs that can be used. We recommend that CNTs be used at much less than

1 ng/cell and that cell viability and wellbeing be followed carefully with all new forms of CNTs and CNT-containing nanodevices. It is likely that CNT toxicity will depend on many other factors than concentration, including their physical form, their diameter, their length, and the nature of attached molecules or nanomaterials.

7 Separation of Fluorescent Material from Single Wall Carbon Nanotubes

For biotechnological uses [39], a high level of purity is required to avoid undesired toxic effects from impurities. Contaminants in SWNT can be classified as carbonaceous (amorphous carbon and graphitic nanoparticles) and metallic (typically transition metal catalysts). It is well documented that nickel, which in combination with yttrium is used as a catalyst in the production of arc-discharged nanotubes, is cytotoxic [40]. Common SWNT purification methods based on oxidation (nitric acid and/or air) have the potential disadvantage of modifying the CNT by introducing functional groups and defects. Other less rigorous purification techniques rely upon filtration, centrifugation and chromatography. Recently, electrophoresis of nitric acid-treated arc-discharged SWNT was used to separate tubular carbon from fluorescent nanoparticles [41].

As we reported in [42], fluorescent nanoparticles were isolated from both pristine and nitric acid-oxidized commercially available carbon nanotubes that had been produced by an electric arc method. The pristine and oxidized carbon nanotube-derived fluorescent nanoparticles exhibited a molecular-weight-dependent photoluminescence in the violet-blue and blue to yellowish-green ranges, respectively. The molecular weight dependency of the photoluminescence was strongly related to the specific supplier. We analyzed the composition and morphology of the fluorescent nanoparticles derived from pristine and oxidized nanotubes from one supplier. We found that the isolated fluorescent materials were mainly composed of calcium and zinc. Moreover, the pristine carbon nanotube-derived fluorescent nanoparticles were hydrophobic and had a narrow distribution of maximal lateral dimension. In contrast, the oxidized carbon nanotube-derived fluorescent nanoparticles were superficially oxidized and/or coated by a thin carbon layer, had the ability to aggregate when dispersed in water, and exhibited a broader distribution of maximal lateral dimension.

The first sample we treated (pNT) was composed of as-prepared nanotubes (AP-SWNT), purified by air oxidation, and the second one (oxNT) was obtained by nitric acid oxidation of AP-SWNT. Both samples had a carbon content in the 80–90 wt-% range and approximately 10 wt-% nickel/yttrium catalyst (4:1). Samples of graphite, carbon black, pNT and oxNT samples were dispersed in aqueous sodium dodecyl sulfate (SDS, 1 wt-%) surfactant using an ultrasonic bath.

Our spectral results reported in [42] may be explained by the presence of fluorescent particles (FP) with variable dimensions and chemistries in the NT samples. The predominant FP in pNT had a mean mw below 30 kDa and exhibited mainly violet-blue photoluminescence and excitation at 315 nm. In contrast, in the fourth fraction (30 kDa–100 kDa), these spectral properties were weaker. All four fractions exhibited 450-nm and 485-nm photoluminescence on 365-nm excitation. The FP in the pNT sample may originate from a nickel and yttrium-containing catalyst that had been covered by a few thin layers of metal oxide and/or carbide during the synthetic process, as was reported by Martinez et al. [43].

Our results suggest that the presence of the FP contaminants in these samples as aggregates or occlusions on NT walls led to quenching of their photoluminescence. Sonication in the presence of the surfactant SDS was able to enrich the solution in micelle-embedded FP. To support this argument, Martinez et al. found that the material encasing the catalyst accompanying nitric acid-oxidized NT masked the metal core from detection by surface-sensitive techniques such as X-ray photoelectron spectroscopy. Their study may also explain the results of Xu et al., who, despite using energy-dispersive X-ray spectroscopy, were unable to discern any metal in the fluorescent electrophoresis fractions from oxidized SWNT [41].

In our study [42], the FP from the oxNT exhibited more size-dependence in their photoluminescence, which ranged from greenish-blue to orange, than those from the pNT. The oxNT FP were also more hydrophilic, probably because their carbon shells became carboxylated on oxidation, as was reported by Xu et al. In addition, our spectra suggest the presence of a less-abundant FP fraction, which had a mean mw below 3 kDa and 425-nm emission peaks on excitation at 315 nm. On the basis of comparing these spectral properties with those of the FP from pNT, FP from the oxNT also contained fluorescent components derived from the pNT.

The above findings led us to design a new SWNT purification method. Each dispersed NT sample was ultracentrifuged (as described above). The pellet was subjected to three additional rounds of dispersion in vehicle, sonication for 5 minutes and ultracentrifugation. A spectrum indicative of few residual FP was exhibited by oxNT-in-water. The spectra of the pNT-in-SDS and oxNT-in-SDS samples resembled that of graphite to demonstrate the absence of fluorescent contaminants. These results suggest that, in addition to amplifying FP photoluminescence, the surfactant facilitated FP removal. In summary, we now have a simple route consisting of surfactant-assisted dispersal followed by ultracentrifugation for removing FP contaminants from both pristine and nitric acid-treated SWNT.

In summary, we isolated, fractionated by molecular weight and characterized FP from pCNT and oxCNT received from several suppliers. These FP were responsible for the photoluminescence of electric arc-produced CNT in the visible range and were likely composed of impurities that were present in the graphite rods used for the production of the CNT. Spectroscopic analysis of the samples revealed some common supplier-independent features, specifically that the FP derived from the pCNT exhibited a violet-blue photoluminescence, whereas the FP derived from the oxCNT exhibited photoluminescence ranging from blue to yellowish-green. In contrast, the molecular weight dependency for both the pristine and oxidized CNT-derived fractions was strongly related to the specific supplier. This can be explained by differing fabrication processes leading to different physical and chemical aggregation of the impurities present in the graphite rod. We recorded HRTEM images and EDX analysis of the FP isolated from the CNT (Carbon Solutions, Inc.)-derived molecular weight fractions. The FP derived from the pCNT exhibited a narrow range in width, whereas the FP derived from the oxCNT were larger, had a broader width range, and formed hydrophilic aggregates in water. Moreover, EDX analysis of the fractions from the oxCNT-in-water supernate suggested that their FP were superficially oxidized and/or coated by a thin carbon layer.

8 Conclusions

The field of nanoscience has been witnessing a rapid growth in the last decade. Recently, more and more, the attention of the community of nanoscientists has been focusing on technological applications. Nanotechnology has been emerging as an enabling technology, with high potential impact on virtually all fields of mankind activity (industrial, health-related, biomedical, environmental, economy, politics, etc.), yielding high expectations for a solution to the main needs of society, although having to address open issues with respect to its sustainability and compatibility.

The fields of application of the research in nanoscience include aerospace, defense, national security, electronics, biology and medicine. There has been a significant progress in understanding achieved in recent years, both from the theoretical and experimental point of view, along with a strong interest to assess the current state of the art of this fast growing field, stimulating, at the same time, research collaboration among the different stakeholders in the area of nanoscience and the corresponding technological applications, prompting possibly the organization and presentation of joint projects in the near future involving both industry and public research.

In the present article, we focused in particular on the biological and medical fields and described current and possible future developments in nanotechnological applications in such areas. Nanostructured, composite materials for drug delivery, biosensors, diagnostics and tumor therapy were reviewed here as examples. Carbon nanotubes were discussed as a primary example of emerging nanomaterials for many of the abovementioned applications.

Acknowledgements The author acknowledges members of his group for nanotechnology at INFN-LNF and collaborators for their work on the developments discussed in this article.

References

1. S. Iijima, Nature, 1991, 354, 56.
2. M.S. Dresselhaus, G. Dresselhaus, P. Avouris (Eds.), 2001. Carbon nanotubes: Synthesis, structure, properties and applications. Springer, Berlin.
3. http://www.germanyinfo.org/relaunch/info/publications/week/2003/030613/misc2.html

4. R. Hergt, R. Hiergeist, J. Hilger, W.A. Kaiser, Y. Lapatnikov, S. Margel, and U. Richter, J. Magnetism and Magnetic Materials, 2004, 270, 345.
5. A. Ito, F. Matsuoka, H. Honda, and T. Kobayashi, Cancer Immunology Immunotherapy, 2004, 53 (1), 26.
6. D.P. O'Neal, L.R. Hirsch, N.J. Halas, J.D. Payne, and J.L. West, Cancer Letters, 2004, 209, 171.
7. D. Luo, E. Han, N. Belcheva and W.M. Saltzman, J. Controlled Release, 2004, 95, 333.
8. A.K. Salem, P.C. Searson and K.W. Leong, Nat. Mater., 2003, 2, 668.
9. G.F. Paciotti, L. Myer, D. Weinreich, D. Goia, N. Pavel, R.E. McLaughlin and L. Tamarkin, Drug Deliv., 2004, 11, 169.
10. K.S. Soppimath, T.M. Aminabhavi, A.R. Kulkarni and W.E. Rudzinski, J. Controlled Release, 2001, 70, 1.
11. P. Nednoor, M. Capaccio, V.G. Gavalas, M.S. Meier, J.E. Anthony and L.G. Bachas, Bioconjug. Chem., 2004, 15, 12
12. T. Konno, J. Watanabe and K. Ishihara, Biomacromolecules, 2004, 5, 342.
13. X.L. Luo, J.J. Xu, W. Zhao and H.Y. Chen, Biosens. Bioelectron., 2004, 19, 1295
14. S. Hrapovic, Y. Liu, K. B. Male and J.H. Luong, Anal. Chem., 2004, 76, 1083.
15. M.C. Daniel and D. Astruc, Chem. Rev., 2004, 104, 293.
16. W. Tan, K. Wang, X. He, X.J. Zhao, T. Drake, L. Wang and R.P. Bagwe, Med. Res. Rev., 2004, 24, 621
17. S. Santra, P. Zhang, K.Wang,R. Tapec andW. Tan, Anal. Chem., 2001, 73, 4988
18. initial He, K. Wang, W. Tan, B. Liu, X. Lin, C. He, D. Li, S. Huang and J. Li, J. Am. Chem. Soc., 2003, 125, 7168.
19. T. Seeger, Ph. Redlich, N. Grobert, M. Terrones, D.R.M. Walton, H.W. Kroto and M. Rühle, Chem. Phys. Lett., 2001, 339, 41
20. E. Whitsitt and A. R. Barron, Nano Lett., 2003, 3, 775.
21. H. Kim and W. Sigmund, Appl. Phys. Lett., 2002, 81, 2085.
22. J.M. Haremza, M.A. Hahn and T.D. Krauss, Nano Lett., 2002, 2, 1253.
23. S. Ravindran, S. Chaudhary, B. Colburn, M. Ozkan and C. S. Ozkan, Nano Lett., 2003, 3, 447.
24. S. Lee and W. Sigmund, Chem. Commun., 2003, 6, 780.
25. J. Sun, L. Gao and M. Iwasa, Chem. Commun., 2004, 7, 832.
26. Massimo Bottini, Lutz Tautz, Huong Huynh, Edvard Monosov, Nunzio Bottini, Marcia I. Dawson, Stefano Belluccib and Tomas Mustelin, Chem. Commun, 2005, 6, 758.
27. J. Liu, A.G. Rinzler, H. Dai, J.H. Hafner, R.K. Bradley, P.J. Boul, A. Lu, T. Iverson, K. Shelimov, C.B. Huffman, F. Rodriguez-Macias, Y. Shon,T.R.Lee, D.T. Colbert and R.E. Smalley, Science, 1998, 280, 1253.
28. J. Esquena, Th. F. Tadros, K. Kostarelos and C. Solans, Langmuir, 1997, 13, 6400
29. F.J. Arriagada and K. Osseo-Asare, J. Colloid Interface Sci., 1999, 211, 210.
30. R.P. Bagwe, C. Yang, L.R. Hilliard and W. Tan, Langmuir, 2004, 20, 8336.
31. H.H. Yang, S.Q. Zhang, X.L. Chen, Z.X. Zhuang, J.G. Xu and X.R. Wang, Anal. Chem., 2004, 76, 1316.
32. G. Fiandaca, E. Vitrano and A. Cupane, Biopolymers, 2004, 74, 55.
33. Massimo Bottini, Andrea Magrini, Almerinda Di Venere, Stefano Bellucci, Marcia I. Dawson, Nicola Rosato, Antonio Bergamaschi, and Tomas Mustelin, J. Nanosci. Nanotechnol., 2006, 6, 1381.
34. Cui, D., Tian, F., Ozkan, C.S., Wang, M., Gao, H., Toxicol. Lett., 2005, 155, 73
35. Monteiro-Riviere, N.A., Nemanich, R.J., Inman, A.O., Wang, Y.Y., Riviere, J.E., Toxicol. Lett., 2005, 155, 377.
36. Liu, J., Rinzler, A.G., Dai, H., Hafner, J.H., Bradley, R.K., Boul, P.J., Lu, A., Iverson, T., Shelimov, K., Huffman, C.B., Rodriguez-Macias, F., Shon, Y.S., Lee, T.R., Colbert, D.T., Smalley, R.E., Science, 1998, 280, 1253.
37. Pantarotto, D., Briand, J.P., Prato, M., Bianco, A., Chem. Commun., 2004, 16 Shi Kam, N.W., Jessop, T.C., Wender, P.A., Dai, H., J. Am. Chem. Soc., 2004, 126, 6850.
38. Bottini M, Bruckner S, Nika K, Bottini N, Bellucci S, Magrini A, Bergamaschi A, Mustelin T., Toxicol Lett., 2006, 160, 121.

39. Baughman RH, Zakhidov AA, de Heer WA., Science, 2002, 297, 787.
40. Pulido MD, Parrish AR., Mutat. Res., 2003, 533, 227.
41. Xu X, Ray R, Gu Y, Ploehn HJ, Gearheart L, Raker K, Scrivens WA, J. Am. Chem. Soc., 2004, 126, 12736.
42. Massimo Bottini, Chidambara Balasubramanian, Marcia I Dawson, Antonio Bergamaschi, Stefano Bellucci, Tomas Mustelin, J Phys Chem B Condens Matter Mater Surf Interfaces Biophys., 2006, 110, 831
43. Martinez MT, Callejas MA, Benito AM., Maser WK, Cochet M, Andres JM, Schreiber J, Chauvet O, Fierro JL., Chem. Commun, 2002, 7(9), 1000.

New Advances in Cell Adhesion Technology

Santina Carnazza

1 Introduction

The main topic of this tutorial is bioadhesion, in terms of both fundamental and applied implications.

First of all, we look how cells adhere to a surface and what are the mechanisms underlying adhesion of both eukaryotic and microbial cells, focusing attention mainly on the cell response to abiotic surfaces.

Then, this paper will review the most recent biotechnological applications requiring the production of hybrid systems through controlled adhesion of biological components (amino acids, peptides, proteins, whole cells) onto polymers and inorganic surfaces. Biomaterials are requested with both good mechanical properties and biocompatibility.

Special attention will be directed to the spatial controlled adhesion, very important in nanotechnology and bioengineering, focusing on methods and application fields. First biomedical applications, and particularly regenerative medicine (including tissue engineering), will be analyzed, in which biomaterials act as *passive* physical surfaces and simultaneously as *active* substrate for cell adhesion, migration, proliferation and differentiation. Most currently developed materials need to evoke cell adhesion and spreading, while potentially preventing bacterial colonization because bacterial adhesion to human tissues and biomaterial surface of biomedical devices is a crucial stage in infection pathogenesis.

And, of course, spatially controlled cell adhesion is requested in BioMEMS applications, in particular for development of biosensors and diagnostic microsystems. Lab-on-chips and microarrays currently used will be reviewed.

The main trends in the BioMEMS research are miniaturization and integration of components and the use of microtechniques to improve immobilization and spatial confinement methods. These and other applications requiring the cell/surface interaction account for considerable efforts in development of surface modification and cellular patterning methods, that are very important tools for fundamental studies in

Santina Carnazza
Department of Microbiological Genetic and Molecular Sciences, University of Messina, Sal. Sperone 31, Vill. S. Agata, 98166 Messina, Italy, e-mail: santina.carnazza@unime.it

S. Bellucci (ed.), *Nanoparticles and Nanodevices in Biological Applications*. Lecture Notes in Nanoscale Science and Technology 7,

biology, especially on single cells, as well as for preparation of chip-based systems in biotechnology. Here, the main cell adhesion technologies will be discussed, and recent progress based on our research results will be briefly reported.

In our laboratory, the biology of fundamental interactions between cells and materials is studied, in relation to the physico-chemical properties of the biomaterial surface. We study cell adhesion in a controlled fashion, using adhesion-supporting and -inhibiting substrata, and analyzing the subsequent cell responses. Additionally, we prepare high resolution micropatterned surfaces for the creation of organized mammalian cell patterns for applications such as biosensors and in particular *single-cell arrays*.

New experimental data will be presented on bio-functionalization of polymer surfaces by controlled ion implantation and fibronectin adsorption aimed to enhance cell adhesion and spatial confinement. Moreover, a new technology will be proposed as an useful tool for preparation of microbial arrays that hold promise as platforms for whole-cell biosensors and diagnostic chips. Another important application for microbial arrays can be in microbial fuel cells, where there is the need for a technology that can provide, in a cost-effective manner, the large surface areas needed for the anodes and cathodes. On the other hand, the ability to obtain ordered microbial arrays with a fractal geometry could overcome problems of blocking and flux control and allow microbial biofilter use for liquid decontamination.

Finally, perspectives are presented of surface bio-functionalization by phage-displayed peptides, which can act as highly specific and selective probes in bioaffinity sensors, can be used in development of nanomaterials and cantilever-based nanodevices for biosensing, and can mimic ligands of cell receptors involved in signaling that affect the cellular fate.

2 A Journey in Cell Adhesion

2.1 Eukaryotic Cell Adhesion

Adhesion is a dynamic interaction between cells and their microenvironment through an extra-cellular matrix (ECM). This matrix is mainly composed of fibrous proteins (collagen, fibronectin, laminin...), glycosaminoglycans (hyaluronic acid, heparin) and proteoglycans.

The association of cells with ECM initiates the assembly of specific cell-matrix adhesion sites. These sites are involved in physical attachment of cells to external surfaces, which is essential for cell migration and tissue formation as well as for activation of adhesion-mediated signaling events.

Cells interact with extracellular matrix primarily through integrins, a widely expressed family of cell surface receptors [60, 129, 130, 220], and integrin binding to its extracellular ligand is responsible for the downstream effects of the matrix on cell function [136].

Integrins are obligate heterodimers containing two distinct chains, called the α (alpha) and β (beta) subunits. In mammals, 19 α and 8 β subunits have been characterized. Through different combinations of these alpha and beta subunits, some 24 unique integrins are generated, although the number varies according to different studies. The integrin binding with its ECM ligand (fibronectin or collagen) triggers signals that involve both physical and biochemical components, including cytosol proteins (talin, paxillin), cytoskeleton (actin and tubulin), tyrosine phosphorylation and second messengers, finally acting as determinants of cell functions.

Following association with their ligands, integrins induce reorganization of the actin cytoskeleton and associated proteins, resulting in the formation of cell-matrix adhesion sites. The best known class of matrix adhesions are the focal contacts (FCs), which contain a multitude of anchor and cytoskeletal molecules such as vinculin, paxillin and talin [39, 133, 275] as well as signal transduction molecules, including different protein kinases, their substrates and various adapter proteins [108, 133, 137, 275, 276]. The recruitment of these molecules to FC and their activation plays a central role in the generation of adhesion signals that are involved in the regulation of many cellular processes such as cell growth, differentiation or apoptosis [60, 220, 274, 275].

The specific type of integrin present in matrix adhesions can vary, depending on the nature of the underlying ECM. The dominant integrin in mature FCs is a $\alpha_V\beta_3$. In addition, fibroblasts can form a distinct class of adhesive contacts in which cell surface integrins bind to fibronectin fibrils in fibrillar adhesions [55, 56, 228]. The process of classical FC assembly may reflect only one type of association of integrins with the ECM, and different cells can display different patterns of matrix adhesions. In typical adherent cells, a large number (tens to hundreds) of distinct FCs can be detected, ranging in size from less than a square micron to several square microns. The number of FCs, their size and distribution, can vary greatly from one cell to the other or even within a single cell, and their morphological diversity may be affected by multiple factors including the nature of the substrate, composition of the medium, incubation time and cell density.

It was demonstrated that the physical state of the ECM, not just its composition, plays a critical role in the regulation of differential assembly of adhesion sites [139]. In fact, the physical properties of the ECM may provide important regulatory signals governing the shape and molecular composition of adhesion sites in a variety of physiological states. Integrins can provide cells with a mechanism to explore and respond to the physical state of the ECM.

Adhesion is influenced by a number of factors: environment (temperature, proliferation time, cell number, growth factors, chemicals, etc.), ECM composition, surface properties (chemistry, roughness, wettability, etc.), specific adhesion (adhesion sites, integrins, focal contacts, vinculins, etc.).

2.2 Microbial Adhesion

Microbial adhesion on natural and abiotic surfaces have pervasive importance in many different aspects of nature and human life, from marine science, soil and plant ecology to food industry, biomedicine and biotechnology (production, wastewater treatment, biosensors) [87]. In these latter, a special attention is directed to mechanisms and methods of microbial patterning that will be discussed later.

In recent years, the interest in microbial adhesion is growing rapidly, since microorganisms have a strong tendency to deposit and adhere on solid surfaces as the onset of a multi-step process, leading to a complex, adhering microbial community termed biofilm [50]. Biofilm processes are manifest in many forms and are studied in a wide range of scientific disciplines. The importance of these microbial communities is twofold: On the one hand, biofilms serve beneficial purposes in the natural environment and in some modulated or engineered biological systems, for example, in the process of degradation and removal of hazardous substances in soil and natural streams, or in a bioreactor or as bioflocculants in wastewater treatment plants; however, on the other hand, the presence of biofilms poses serious problems, for instance on food, on ship hulls, on old-fashioned portraits, on historical monuments and most importantly in biomedical field. Microbial adhesion to human tissue surfaces and implanted biomaterial surfaces is an important step in the pathogenesis of infection. Bacterial infection is the most common cause of biomaterial implant failure in modern medicine [80]; as a frequent complication, in fact, adhesion and subsequent surface growth of bacteria on biomedical implants and devices causes the formation of a biofilm [37]. Accordingly, considerable research efforts are directed toward the understanding and control of the cascade of events leading to the bacterial colonization of synthetic surfaces in the form of biofilm [93,224,243,262,271]). Such a process is commonly discussed in terms of the two-phase sorption model first suggested by Marshall et al. [170]. In this model, the bacteria during the first step are rapidly attracted to a given surface by weak physical interactions, producing a basically reversible attachment, while in the second step, an irreversible molecular and cellular adhesion process occurs, and aggregates resistant to any washing treatment are formed. More recently [215], the model has been upgraded by including five stages of biofilm development: (1) reversible attachment: cells are weakly attached to substrate; (2) irreversible attachment: cells are cemented to the substrate and form nascent cell clusters; (3) maturation-1: cell clusters mature and become progressively layered, embedded in a matrix of extracellular polymeric substances (EPS); this stage is accompanied by the activation of quorum-sensing signaling; (4) maturation-2: after several days of growth, cell clusters reach their maximum thickness; and (5) dispersion: motile cells swim away from the inner portions of cell clusters and enter the bulk liquid, altering the biofilm structure by forming void spaces.

In particular, during the first few hours after the initial attachment of motile bacteria to a solid support, the bacteria aggregate into microcolonies, as a consequence of individual cell twitching across the surface [185]. Thus, a few adhering sessile microorganisms, depending on the substrate surface properties, can stimulate the adhesion of other planktonic bacteria, producing the so-called co-adhesion phenomenon [27]. This process, in turn, is responsible for the formation of a linking film, determining the characteristic features of the biofilm as a whole, in terms of both structure and adhesion properties [40].

It has been also demonstrated that the different phases of biofilm development can be correlated with specific physiological characteristics, involving distinct structural and metabolic changes [215]. In particular, the irreversible adhesion onto a given surface has been shown to occur through the production of EPS, which is often stimulated by the attachment process itself [2, 255].

From an overall physico-chemical point of view, the initial, instantaneous phase of microbial adhesion is mediated by nonspecific interactions, with long-range characteristics, including Lifshitz-van der Waals forces, electrostatic forces, acid-base interactions, hydrophobic interactions, and Brownian motion forces, depending on a relatively limited number of physico-chemical properties of the surfaces, generally treated in the framework of the Derjaguin–Landau–Verwey–Overbeek (DLVO) theory [28, 181]. Marshall et al. [170], for the first time, suggested that the initial reversible phase could be explained in terms of the DLVO theory of colloid stability. Since then, the classical DLVO theory has been used by many workers as a qualitative model, but also in some cases in a quantitative way to actually calculate adhesion free energy changes in order to explain microbial adhesion. In short, the DLVO theory has been used to describe the net interaction between a cell and a flat solid surface, as a balance between attractive and repulsive interactions. This theory assumes cells to be perfectly smooth spherical surfaces which, in reality, do not exist. Such considerations lead to treat the microorganisms as living colloids, disregarding the specific roles of bacterial structures/molecules such as pili, cell wall components, capsule material and lipopolysaccharides (LPS), which are considered to be of particular importance in the later stages of adhesion and particularly in biofilm formation. In the present state of art, it appears clear that the critical step for the formation of the biofilm is the irreversible attachment step, i.e., the second step in the framework of the Marshall model [170]. As suggested above, this step seems to be critically related to specific surface macromolecules responsible for adhesive activities of microorganisms, the so-called adhesins (for a concise review, [3]). Bacteria may have multiple adhesins for different substrata, commonly lectins or lectin-like proteins or carbohydrates, parts of surface polymeric structures, which include capsules, fimbriae or pili, and slime. However, it has been argued that these structural features are of less significance in the initial stages of the attachment process than the intrinsic thermodynamic factors involved [3, 179], and a number of detailed studies have been carried out to support this assertion [72, 253]. Bacterial adhesion to many materials has been successfully described in terms of colloidal interaction forces governed by physicochemical properties of the bacteria and the solid surfaces (for an overview [28]). This was in contrast to the thermodynamic approach, which often produced results inadequate for a general description of bacterial adhesion. In the DLVO theory, the interaction energy is distance dependent, whereas in the thermodynamic approach, the formation of a new cell-substratum interface at the expense of the substratum-medium interface is calculated, i.e. the strength of the interaction at contact is achieved [123]. However, there are a number of experimental cases where the thermodynamic approach does not adequately explain the results. The extended DLVO theory, suggested by van Oss [254], uses components from both models, and includes acid-base interactions (hydrophobicity/hydration effects), in addition to the classical van der Waals (vdW) and electrostatic interactions. In some cases, the extended DLVO theory qualitatively predicts experimental adhesion results better than the classical DLVO theory and the thermodynamic approach, but the strength of the adhesion energy seems to be overestimated, possibly because only a very small fraction of the cell is actually in direct contact with the substratum.

It is worth noting that a satisfactory theory for microbial adhesion is up till now lacking, mainly because a physico-chemical approach will most likely never be able to fully explain all aspects of microbial adhesion to surfaces, including interspecies binding. In adhesion experiments, most often, only one microbial strain is investigated at a time. This is not relevant for natural systems, where a surface is exposed to a whole range of different microbial species that attach simultaneously and sequentially. Adhesion of one cell is likely to be affected by the presence of other cells at the surface, so called co-adhesion. Finally, some biological factors may be of importance and can hardly or ever be accounted for in theoretical models described above. The attached cells can *sense* the surface and develop adaptive traits; if surfaces are *sensed* by microorganisms, phenotypic changes may occur very quickly. This, in turn, makes adhesion predictions difficult by present physico-chemical models.

In conclusion, as for eukaryotic cells, microbial adhesion is a very complicated process that is affected by many factors, including some properties of the cell itself (hydrophobicity, surface charge, specific adhesins, virulence factors, etc.), the target material surface (chemical composition, roughness, wettability, etc.), and the environmental factors (temperature, time period of exposure, cell number, presence of antibiotics, chemicals, etc.).

3 Biotechnological Applications

The most recent biotechnological applications in fields like biocompatible surfaces, bioelectronics, biosensors etc. require the production of hybrid systems based on the controlled adhesion of biological systems, going from amino acids and peptides till proteins and whole cells, onto polymers and inorganic surfaces [239]. Biomaterials are requested with both good mechanical properties and biocompatibility [82].

Biomaterials is a term used to indicate materials, synthetic or natural in origin, which can be used as a whole or as a part of a system which treats, augments or replaces any tissue, organ or function of the body [26]. Synthetic materials currently used for biomedical applications include metals and alloys, composites and polymers, ceramics and bioglass. Because the structures of these materials differ, they have different properties and, therefore, different uses. Synthetic polymers are the most widely used biomaterials since they are the most favorable for their various composition, properties and shape and the possibility of prompt fabrication in complex shapes and structures. They are the materials of choice for cardiovascular devices as well as for replacement and augmentation of various soft tissues. Polymers also are used in drug delivery systems, in diagnostic aids, and as a scaffolding material for tissue engineering applications.

Biomaterials interact with the body through their surfaces. Consequently, the properties of the outermost layers of a material are critically important in determining both biological responses to implants and material responses to the physiological environment. Changes in surface characteristics further modify biological responses. A goal of biomaterials research is to design surfaces that elicit desired interfacial behaviors in order to control biological responses. Biomaterial surface engineering offers the ability to modify material and biological responses through changes in surface properties while maintaining the bulk properties. Biomaterial surface engineering approaches can be classified according to the surface properties being altered, e.g., physicochemical, morphologic, or biological modifications.

It is to be noted that most relevant biotechnological applications need a spatially resolved process of adsorption/adhesion of the biological systems of interest. Special attention is therefore directed to the spatial controlled adhesion methods, very important in bioengineering and nanotechnology.

In the micrometer range, the development of highly integrated multipurpose microsystems, including biosensor arrays, DNA-chips and related technologies, as well as the lab-on-chip devices critically depend on the ability to functionalize specific micrometer-scale areas in very complex three-dimensional structures, mostly based on silicon and related compounds [4].

On the other hand, at the nanometer scale, at least two different aspects of controlled biomolecule adsorption/organization must be considered, i.e., the interest in achieving nanosized patterns of biological molecules in a general perspective of bioelectronics devices, and the specificity of nanostructured surfaces in stimulating specific response of large systems, as cells or protein aggregates, opening new, very interesting views on the way the cells can *sense* nanometric surface features [77, 187, 268].

In view of the above sketched areas of application, suitable and application-specific methods and techniques are developing in the current research to obtain the spatially-resolved structuring of surface.

4 Why Controlled Cell Adhesion?

Let us discuss in some detail the application fields for controlled cell adhesion. First of all, biomedicine, particularly regenerative medicine (including tissue engineering), in which biomaterials not only anchor cells as *passive* physical surfaces, supporting cell spreading and migration, but simultaneously act as *active* substrate, directly involved in modulation of the main processes of cell life: survival, proliferation and expression of specific developmental phenotypes such as differentiation and apoptosis. Most currently developed materials need to evoke cell adhesion and spreading, but both tissue cell integration and microbial adhesion can be enhanced, since both of them have interrelated and similar mechanisms and depend largely on the surface structure and composition of implanted biomaterials. Thus, a *race for the surface* appears, which is often won by microorganism causing colonization and subsequent infectious complications of the medical device. Considerable research efforts are currently directed towards the development of polymeric materials potentially preventing bacterial colonization. A variety of methods are utilized in an attempt to reduce the incidence of bacterial adhesion or, at least, to lower the number of viable bacteria. One of the strategies that have emerged as potential means of inhibiting the early stages of bacterial colonization involves the surface modification of polymeric materials which function by creating a non-stick surface to colonizing microorganism. Last, but not least, spatially controlled cell adhesion is requested in BioMEMS applications, in particular for development of biosensors and diagnostic microsystems.

4.1 Tissue Engineering

Tissue engineering for the realization of parts of or whole artificial organs is a very important and challenging area of research.

It is a multidisciplinary field which involves the application of the principles and methods of engineering and life sciences and the development of biological substitutes that restore, maintain or improve tissue function [222].

One of the main methods of tissue engineering involves growing cells in vitro into the required three-dimensional organ or tissue. But cells lack the ability to grow in 3D orientations: they randomly form a two-dimensional layer. However, 3D tissues are required and this is achieved by seeding the cells onto porous matrices, known as scaffolds, to which the cells attach and colonize [150]. These cells then proliferate, migrate and differentiate into the specific tissue while secreting the extracellular matrix components required to create the tissue.

The scaffold therefore is a very important component for tissue engineering. To achieve the goal of tissue reconstruction, scaffolds must meet some specific requirements [128]. A high porosity and an adequate pore size are necessary to facilitate cell seeding and diffusion throughout the whole structure of both cells and nutrients. Biodegradability is essential since scaffolds need to be absorbed by the surrounding tissues without the necessity of a surgical removal. The rate at which degradation occurs has to coincide as much as possible with the rate of tissue formation: this means that while cells are fabricating their own natural matrix structure around themselves, the scaffold is able to provide structural integrity within the body and eventually it will break down, leaving the neotissue, newly formed tissue, which will take over the mechanical load. Injectability is also important for clinical uses; in any case, it should not induce any adverse host response. The scaffold should be made from material with appropriate surface chemistry to favor cellular attachment, differentiation and proliferation, possess adequate mechanical properties to match the intended site of implantation and handling, and be easily fabricated into a variety of shapes and sizes.

Many different materials (natural and synthetic, biodegradable and permanent) have been investigated. Most of these materials have been known in the medical field before the advent of tissue engineering as a research topic, being already employed as bioresorbable sutures, e.g. collagen or some linear aliphatic polyesters.

New biomaterials have been engineered to have ideal properties and functional customization: injectability, synthetic manufacture, biocompatibility, non-immunogenicity, transparency, nanoscale fibers, low concentration, resorption rates, etc. A commonly used synthetic material is polylactic acid (PLA). This is a polyester which degrades within the human body to form lactic acid, a naturally occurring chemical which is easily removed from the body. Similar materials are polyglycolic acid (PGA) and polycaprolactone (PCL): their degradation mechanism is similar to that of PLA, but they exhibit respectively a faster and a slower rate of degradation compared to PLA.

Scaffolds may also be constructed from natural materials: in particular, different derivatives of the extracellular matrix have been studied to evaluate their ability to support cell growth. Proteic materials, such as collagen or fibrin, and polysaccharide materials, like chitosan or glycosaminoglycans (GAGs), have all proved suitable in terms of cell compatibility, but some issues with potential immunogenicity still remains. Among GAGs, hyaluronic acid, possibly in combination with cross linking agents (e.g. glutaraldehyde, water soluble carbodiimide, etc.), is one of the possible choices as scaffold material. Functionalized groups of scaffolds may be useful in the delivery of small molecules (drugs) to specific tissues.

For an overview of the most commonly used scaffold materials and conventional and new scaffold fabrication techniques, see Sachlos and Czernuszka [202].

4.2 BioMEMS

BioMEMS (Biological Micro-Electro-Mechanical Systems) are devices constructed using techniques inspired from micro/nanoscale fabrication, that are used for processing, delivery, manipulation, analysis, or construction of biological and chemical entities [14].

In recent years, the biological and biomedical applications of micro- and nanotechnology have become increasingly prevalent, and BioMEMS have found widespread use in a wide variety of applications that encompass all interfaces of biology and biomedical sciences with micro- and nanoscale systems. Areas of research and applications in BioMEMS range from diagnostics to systems for drug delivery, novel materials for Bio-MEMS, microfluidics, tissue engineering, surface modification, etc.

In their simplest form, technologies in the BioMEMS arena leverage advances in micro-fabrication and micro-machining to create faster, cheaper, hands-off micro- and nanoscale laboratories. In more sophisticated forms, BioMEMS devices offer an avenue to artificial organs, personalized drug therapies and new ways to view cell communication. BioMEMS can be characterized into two categories: biomedical MEMS, which deal in vivo with the body and the host anatomy, and biotechnological MEMS, which deal in vitro with the biological samples from the host. Examples of biomedical MEMS include minimally-invasive therapy precision surgery, biotelemetry, drug delivery, biosensors and other physical sensors. Examples of biotech MEMS include gene sequencing, functional genomics, drug discovery, pharmacogenomics, diagnostics, and pathogen detection.

Diagnostics represents the largest and most researched BioMEMS area. A very large and increasing number of BioMEMS devices for diagnostic applications, referred to as *BioChips*, have been developed within the last few years. These devices are used to detect cells, microorganisms, viruses, proteins, DNA and related nucleic acids, and small molecules of biochemical interest. They are essentially miniaturized laboratories that can perform hundreds or thousands of simultaneous biochemical reactions. In general, the use of micro- and nanoscale detection technologies is justified by (1) reduced scale of the sensor element, providing a higher sensitivity; (2) reduced reagent volumes and associated costs; (3) reduced time to result due to small volumes resulting in higher effective concentrations; and (4) amenability of portability and miniaturization of the entire system.

BioMEMS hold a lot of promise also for the analysis of *single cells* and the study of their function in *real time*. Micro- and nanoscale systems and sensors could allow us to precisely measure the protein, mRNA, and chemical profiles of cells in real time, as a function of controlled stimulus and increase understanding of signaling pathways inside the cell. These are essential to increase our understanding of the underlying cause of basic cell functions such as differentiation, reproduction, apoptosis, etc., and their implications on various disease states. To accomplish these goals, BioMEMS can play an important role, especially in the development of integrated devices and systems for the rapid and real-time analysis of cellular components, especially from single cells. The development of microenvironments where cells can be precisely placed, manipulated, lysed, and then analyzed using micro- and nano-sensors in real time, would have a significant impact on *systems biology*. Integration of sensors for detection of DNA, mRNA, proteins, and other parameters indicating cellular conditions such as oxygen, pH, etc., can be accomplished using BioMEMS platforms and nanoscale sensors.

The devices and integrated systems using BioMEMS are also known as lab-on-a-chip and micro-total analysis systems (micro-TAS or μTAS). The word is now used very broadly and applied to devices with no electro-mechanical components, such as DNA and protein arrays. Some selected examples of BioMEMS will be introduced below, and the biological aspects of nanotechnology applications will be mainly described.

4.2.1 Biosensors

Biosensors are analytical devices that combine a biologically sensitive element with a physical or chemical transducer to selectively and quantitatively detect the presence of specific compounds in a given external environment [257].

The biological element can be a single enzyme, nucleic acid, antibody or a whole microorganism or also a mammalian or plant tissue. Detection systems can be mechanical, optical, electrical.

Commercially available are enzymatic biosensors for glucose, lactate, alcohol, sucrose, galactose, uric acid, alpha amylase, choline and L-lysine. All are amperometric sensors based on O_2 consumption or H_2O_2 production in conjunction with the turnover of an enzyme in the presence of substrate. However, there is currently a wide interest in using whole cells as functional components in biosensors for their built-in natural selectivity to biologically active chemicals and the physiologically relevant response [29, 188, 234]. The transduction of the cell sensor signals may be achieved by the measurement of transmembrane and cellular potentials, impedance changes, metabolic activity, analyte inducible emission of genetically engineered reporter signals, and optically, by means of fluorescence or luminescence.

Many mammalian cell-based bioassays for the screening of large libraries of potential pharmaceutical agents or the detection of toxic compounds and pathogens in the environment have been developed over the years and some are in use today. With assay miniaturization in mind, a number of researchers have explored the patterned deposition of cells in microsystems through the control of cellular adhesion.

Similarly, the control of bacterial adhesion may be a valuable tool toward developing cell-based sensor arrays based on genetically engineered bacteria. Since analyte specificity can be readily modified by genetic engineering and because of the relatively robust nature of these microorganisms as compared to mammalian cells, bacterial cells have been studied extensively for sensing applications [17,157]. Bacterial genome can be altered using recombinant DNA technology and microorganism can be constructed, potentially, to harness energy, decompose toxic waste, and possibly perform computational functions. As the field progresses, there will be a need for tools and technologies to perform gene insertions into single or very few bacteria, to specifically manipulate their characteristics within a network of bacteria. The tools and platforms to perform such integrated synthetic biology can be provided by BioMEMS and related nanoscale sensors, processing, and device technologies.

The ability to selectively attach bacteria to micropatterned substrates could be used to create sensor arrays for applications such as rapid screening for infectious diseases or to detect toxic compounds [68]. A number of studies reported that cellular adhesion is controlled by the chemical and physical characteristics of surfaces such as hydrophobicity and hydrophilicity, surface charge, and surface roughness. These factors can be readily manipulated to control the spatial distribution of cells on a substrate. The rapid development of micro-fabrication and surface engineering techniques have stimulated the development of novel methods that may be used to control cell adhesion and spreading with micrometer-scale resolution.

Biosensors find extensive applications in environmental pollution control for measuring toxic gases in the atmosphere and toxic soluble compounds in water. These pollutants may include heavy metals, nitrates, nitrites, herbicides, pesticides, polyaromatic hydrocarbons, polychlorinated biphenyls, etc. The estimation of organic compounds is very important also for the control of food manufacturing process and for the evaluation of food quality. The online analysis of raw materials and products is also necessary in industrial fermentation processes. However, a survey of the sensor market identifies medical diagnostics as a major application field for emerging biosensors. Over the years, a number of sensors have been developed for analytes, either as sensor arrays designed for multiple analytes or for specific analytes in specific sensing environments. Biomedical micro- and nanodevices have potential uses that range from the analysis of biomolecules to disease diagnosis, prevention and treatment. The possibility of improved speed, greater sensitivity, reduced cost and decreased invasiveness, has generated substantial interest in miniaturized devices.

4.2.2 High-Throughput Screening

High-throughput screening (HTS) is an automated and miniaturized process by which large numbers of compounds are tested for activity as inhibitors or activators of a particular biological target, such as a cell surface receptor or a metabolic enzyme. Today, most pharmaceutical companies use HTS as the primary tool for early stage drug discovery.

The more steps required for an assay, the more difficult to automate the HTS. The ideal assay is one that can be performed in a single well with no other manipulation other than addition of the sample to be tested.

A number of assay formats have been developed or modified over the past few years; these can be divided into two groups [227]: cell-free assays that measure the biological activity of a relatively pure protein target and cell-based assays that assess the activity of a target protein, by monitoring the biological response of a cell in which the target protein resides. Most recently, cell-based assays are an increasingly attractive alternative to in vitro biochemical assays for HTS, because they have significant advantages over in vitro assays. First, the starting material (the cell) self-replicates, avoiding the costs for preparing a purified target, chemically modifying the target to suit the screen, and so on. Second, targets and readouts are examined in a biological context that more faithfully mimics the normal physiological situation. Third, cell-based assays can provide insights into bioavailability and cytotoxicity, the use of reporter genes, as shown before. However, mammalian cells are expensive to culture and difficult to propagate in the automated systems used for HTS, so an alternative is to reproduce the desired human physiological process in a microorganism such as yeast [144]. It is relatively easy to transfect yeast with human DNA encoding receptors or other components of signal transduction pathways, and this fact has been used to develop a HTS screening system [30, 247]. The signaling pathways in yeast and humans are sufficiently related to permit functional evaluation of human receptors in yeast cells [119, 143, 147, 193, 217]. Because of the straightforward manipulation of genetic information in yeast, it is relatively simple to generate a family of strains that differ by a single human gene, to facilitate the analysis of screening data [241, 282]. The ease and low cost of growing yeasts, their ready genetic manipulation, and their resistance to solvents make yeast an attractive option for cell-based HTS. A variety of assay technologies continue to be developed for HTS, including cell-based assays, surrogate systems using microbial cells such as yeast and bacterial two-hybrid and three-hybrid systems [100, 124].

Over the years, screening of chemical libraries of increasing size has widely dominated the approach taken by the screening community. Emphasis has recently been placed on *in silico* methods for rational selection of drug libraries for the respective targets in screening and novel assay technologies, which will either optimize the performance and throughput or allow the screening of novel biological targets resulting from ongoing genomic/proteomic activities [96].

4.2.3 Lab-on-a-Chip

Lab-on-a-chip is a term used to describe sensors and devices with some level of integration of different functions and functionality. They can include more than one step of analysis, for example, sample preparation and detection, cell lysing and PCR, cell growth and detection of metabolites, etc. Various modules could be used in appropriate combination for the detection of desired entity.

The use of glass and silicon micromachining has become a central tool in development of many types of lab-on-chip sensors. Microfabrication has focused on fluidic channels, optical windows and electrodes with dimensions ranging from millimeters to micrometers. Integrated systems that incorporate both microcapillary electrophoresis and optical or electrical detection form lab-on-chip systems. The strong interest in making highly parallel arrays of capillaries for HTS and DNA diagnostics has prompted intense development activity in microcapillary arrays both in academe and in industry.

Several example of such devices have been reported for the processing and detection of cells, proteins, DNA and small molecules (for a review see [14]).

A possible integrated platform for nanoliter DNA analysis has been described [38]. The device was able to mix solutions, amplify and digest DNA, separate the products by on-chip capillary electrophoresis and detect them by on-chip photodiode fluorescence detectors.

Another very attractive technology has been recently described [164]. A microfluidic device on a CD type platform using centrifugal and capillary forces for liquid transport has been reported. Such devices are very attractive due to their low cost, CD-type format, and integration with available optical detection technology.

In addition, many components of an integrated lab-on-a-chip are under development, including micro-valves, micro-pumps, mixers, selective capture elements, etc.

Many devices are being developed as disposable plastic biochips to prevent cross-contamination. Polymer and hydrogel-based microdevices have many attractive features for use in biomedical lab-on-chip applications such biocompatibility, low cost combined with rapid prototyping techniques, and microfabrication of polymers. The primary use for such micro-total analysis systems has, so far, been in tabletop instruments; however, there is every reason to believe that such devices will soon work their way into in vivo systems as well. Chip-scale, MEMS-base transdermal devices could enable continuous access to body fluids and the continuous monitoring of molecular, cellular and physiological biomarkers. The possible integration of these diagnostic devices for intelligent and integrated sensing with therapeutic micro/nanoscale technologies able to deliver known types and quantities of stimulus, drugs and chemicals, perhaps in a real-time feedback configuration, would be highly beneficial.

4.2.4 Microarrays

Microarrays are critical components of a biochip platform: dense, two-dimensional grids of biosensors typically deposited on a flat substrate, which may be either passive (silicon or glass) or active (consisting of integrated electronics or micromechanical devices) performing or assisting signal transduction.

DNA microarrays are the most successful example of the technology. The techniques used to define patterns on semiconductor surfaces were utilized to construct arrays of single strands of known DNA sequences (capture probes). Once the probes are placed at specific known sites on a chip surface, hybridization with molecules of unknown sequence (target probes) can reveal the sequence. DNA arrays can be formed by two different approaches, i.e. optical [102] and electrical [122]. The detection of hybridization, in both cases, is typically done by fluorescence; however, it can also be done electrically or through other *label-free* detection methods that can result in ease of use, reduced reagents and processing costs, amenability to portability and miniaturization.

Microarray-based genomic technologies have revolutionized genetic analyses of biological systems. They are widely used to monitor gene expression under different cell growth conditions, detect specific mutations in DNA sequences and characterize microorganisms in environmental samples. The widespread, routine use of such genomic technologies will shed light on a wide range of important research questions from how cells grow, differentiate and evolve, to the medical challenges of pathogenesis, antibiotic resistance and cancer; from agricultural issues such as seed breeding and pesticide resistance, to the biotechnological challenges of drug discovery and the remediation of environmental contamination. Several types of microarrays have recently been developed and evaluated for bacterial detection and microbial community analysis. These studies indicated that microarray-based genomic technologies have great potential as specific, sensitive, quantitative, parallel high-throughput tools for microbial detection, identification and characterization in natural environments [280].

Microarrays are not limited to DNA analysis; protein microarrays, antibody microarrays, peptide microarrays, chemical compound microarrays and tissue microarrays can also be produced using biochips.

Protein and antibody arrays can play a key role in search for disease-specific proteins that have medical, diagnostic, prognostic and commercial potential as disease markers or as drug targets and for determination of predisposition to specific disease via genotypic screening [95, 106, 258, 281]. With the recent advancements in genomics and proteomics technologies, many new gene products and proteins are being discovered daily, and array-based chips hold a great potential to systematically analyze the massive amount of data, assigning a biological function to these proteins and determining protein-protein and protein-DNA interactions. The proteins can be arrayed by soft lithography [183] and micro-contact printing [138] to generate protein chips, and each protein spot can be addressed by other proteins to determine recognition events and kinetics. The binding is conventionally detected by fluorescence-based methods, which are safe, sensitive, can have a high resolution, and are compatible with standard microarray scanners, but also immunologically or by other methods.

The most common protein microarray is the antibody microarray, where antibodies are spotted and fixed onto a chip surface (glass, plastic or silicon) and are used as capture molecules for the purpose of detecting antigens. Antibody microarray is often used for detecting protein expressions from cell lysates in general research and special biomarkers from serum or urine for diagnostic applications. Capture molecules used are most commonly monoclonal antibodies; however, more recently there has been a push towards other types of capture molecules which are more similar in their nature such as peptides or aptamers. Antibodies have several problems, including the fact that there are not antibodies for most proteins and also problems with specificity in some commercial antibody preparations.

Peptide chips have been developed consisting of high-density peptide arrays enabling multiple assays of peptide-protein interactions in parallel. Peptide microarrays have been generated in configurable patterns by the high-voltage CMOS technology on microchips as active supports for both combinatorial array synthesis and assay evaluation (http://www.kip.uni-heidelberg.de/ti/PeptideChip). Integrated photodiodes may be used for subsequent detection of interaction patterns of these peptide arrays with labeled proteins. An interface with appropriate software has been developed to allow programming of peptide sequences, experiment control, data analysis and visualization.

A chemical compound microarray is a collection of organic chemical compounds spotted on a solid surface, such as glass and plastic. This microarray format is very similar to DNA microarray, protein microarray and antibody microarray. In chemical genetics research, they are routinely used for searching proteins that bind with specific chemical compounds [125]; and, in general drug discovery research, they are used for searching potential drugs for therapeutic targets [162]. There are three different forms of chemical compound microarrays based on the fabrication method. The first form is to covalently immobilize the organic compounds on the solid surface with diverse linking techniques; this platform is usually called Small Molecule Microarray [250]. The second form is to spot and dry organic compounds on the solid surface without immobilization, this platform has a commercial name as Micro Arrayed Compound Screening (μARCS), which is developed by Abbott Laboratories. The last form is to spot organic compounds in a homogenous solution without immobilization and drying effect, this platform is commercialized as DiscoveryDot™ technology by Reaction Biology Corporation.

Tissue microarrays consist of paraffin blocks in which up to 1000 separate tissue cores are assembled in array fashion to allow simultaneous histological analysis. The technique of tissue microarray was developed to address the major limitations in molecular clinical analysis of tissues, including the cumbersome nature of procedures, limited availability of diagnostic reagents and limited patient sample size. Multi-tissue blocks were first introduced by Battifora in [15] 1986 with his so called *multitumor (sausage) tissue block*. In 1998, Kononen and collaborators [145] developed the current technique, which uses a novel sampling approach to produce tissues of regular size and shape that can be more densely and precisely arrayed. In the tissue microarray technique, a hollow needle is used to remove tissue cores as small as 0.6 mm in diameter from regions of interest in paraffin embedded tissues such as clinical biopsies or tumor samples. These tissue cores are then inserted in a recipient paraffin block in a precisely spaced, array pattern. Sections from this block are cut using a microtome, mounted on a microscope slide and then analyzed by any method of standard histological analysis. Each microarray block can be cut into hundreds of sections, which can be subjected to independent tests. Tests commonly employed in tissue microarray include immuno-histochemistry and fluorescent in situ hybridization. Tissue microarrays are particularly useful in analysis of cancer samples.

Most of the microarrays rely on the labeling of samples with a fluorescent or radioactive tag that is a time-consuming and expensive procedure. Alternative most recently reported label-free methods include nanomechanical biodetection by cantilever microarrays. It requires no labels, optical excitation or external probes, and is rapid, highly specific, sensitive and portable. Cantilever arrays are microfabricated by standard low-cost silicon technology, and can detect DNA hybridization, single bases mismatches, protein-protein, drug-binding interactions and other molecular recognition events in which physical steric factors are important.

4.2.5 Nanomaterials and Nanodevices

The most recent advances in nanotechnology have taken advantage of biochemically induced surface stress to directly and specifically transduce molecular recognition into nanomechanical responses in cantilever arrays. This has been achieved by immobilizing a monolayer of receptor molecules on one side of the cantilevers and then detecting in situ the mechanical bending induced by ligand binding in a liquid environment, using an optical beam deflection technique. A major advantage of such a direct transduction is that it eliminates the requirement that the molecules under investigation be labelled, for example, with fluorescence or radioactive tags. It was first reported the specific transduction, via surface stress changes, of DNA hybridisation and receptor-ligand binding into a direct nanomechanical response of microfabricated cantilevers [105]. Cantilevers in an array were functionalized with a selection of biomolecules, and the differential deflection of the cantilevers was found to provide a true molecular recognition signal despite large nonspecific responses of individual cantilevers. The experiments demonstrate that the differential bending was clearly sequence-specific and provided an unambiguous *yes* or *no* response. Moreover, the method had the intrinsic sensitivity needed to detect single nucleotide polymorphisms, suggesting the capability to sequence DNA by hybridisation or to determine single base mutations, which was demonstrated shortly later [120]. Similar experiments were performed on the specific binding of the constant region of immunoglobulins to protein A, demonstrating the wide-ranging applicability of nanomechanical transduction to detect biomolecular recognition [105]. As an example of both protein–protein binding in general, and tumor marker detection in particular, the application of this technique for sensitive and specific detection of prostate-specific antigen (PSA) has been reported [272].

These studies demonstrate the direct translation of biomolecular recognition into nanomechanics. The method has important advantages in that it does not require labeling, optical excitation, or external probes. Furthermore, the transduction process is repeatable when denaturation or unbinding agents are used, enabling cyclic operation. The methodology is compatible with silicon technology and is suited for in situ operation. Parallelization into integrated devices, recently demonstrated with multiple arrays, allows a new generation of DNA chips and binding assays to be developed based on nanomechanics.

A nanomechanical cantilever array for multiple quantitative biomolecular detection has been developed [174], able to detect multiple unlabeled biomolecules simultaneously, at nanomolar concentrations, within minutes, in a single-step reaction without any sample manipulation. Besides being label free, this technology readily lends itself to formation of microarrays using well-known microfabrication techniques [163] and the possibility to monitor thousands of cantilevers simultaneously with integrated readout, thereby offering the promising prospect of high-throughput nanomechanical genomic analysis, proteomics, biodiagnostics, and combinatorial drug discovery.

The forces involved in these nanosystems are sufficient to operate micromechanical valves and related microfluidic devices. This would also permit the autonomous operation of micro- or nanorobotic machinery. Because the transduction eliminates the need for external control systems, in situ delivery devices could be triggered directly by signals from single cells, gene expression or immune responses.

Advances in nanotechnology and engineering are providing opportunities for use of nanomaterials also in water purification (for an overview, see [216]). Metal-oxide nanoparticles, carbon nanotubes, dendrimers and zeolites are selected nanomaterials currently evaluated as functional materials for water purification. Nanomaterials have a number of key properties that make them particularly attractive as separation media for toxic metal ions, radionuclides, organic and inorganic solutes, bacteria and viruses: in particular, their large surface area and the possibility of functionalization with various chemical groups to increase their affinity toward a given compound. Recent advances suggest that many of the issues involving water quality could be resolved or greatly ameliorated using nanoparticles, nanofiltration or other products resulting from the development of nanotechnology. Innovations in the development of novel technologies to desalinate water are among the most exciting and promising. Utilization of specific nanoparticles either embedded in membranes or on other structural media that can effectively, inexpensively and rapidly render unusable water potable is being explored at a variety of institutions.

Additionally, nanotechnology-derived products that reduce the concentrations of toxic compounds to sub-ppb levels can assist in the attainment of water quality standards and health advisories.

Recent years have witnessed significant interest in biological applications of novel inorganic nanomaterials such as nanocrystals [33, 238], nano-wires [70], and nanotubes [52, 270] with the motivation to create new types of analytical tools for life science and biotechnology.

Carbon nanotubes (CNTs) are interesting molecular wires (diameter $\sim$ 1–2 nm) with unique electronic properties that have been spotlighted for future solid-state nanoelectronics [89, 90]. CNTs exhibit, in general, an unique combination of excellent mechanical, electrical and electrochemical properties [16], which has stimulated increasing interest in their use as components of biosensors for the detection of a number of biomolecules (for a review, [11]). Aligned multi-wall carbon nanotubes (MWNT) grown on platinum substrate have been used for the first time for the development of an amperometric biosensor [232]. The opening and functionalization by oxidation of the nanotube array allowed for the efficient immobilization of the model enzyme, glucose oxidase. The carboxylated open-ends of nanotubes were used for the immobilization of the enzymes, while the platinum substrate provided the direct transduction platform for signal monitoring. It was also shown that carbon nanotubes could play a dual role, both as immobilization matrices and as mediators, allowing for the development of a third generation of biosensor systems, with good overall analytical characteristics. A year later, electrochemical biosensors were constructed by using carbon nanotubes compacted into pellets; alkaline phosphatase was immobilized on the surface of MWNTs utilizing a layer-by-layer methodology [156]. A more recent work demonstrated the use of carbon nanotubes (CNTs) for the development of a low-cost disposable biosensor for the sensitive detection of organophosphorus (OP) pesticides [134]. The dual role of CNT, electrocatalytic activity toward thiocholine and immobilization matrix for the enzyme was demonstrated. This led to the development of a mediator free, simple and robust single enzyme biosensor for the sensitive detection of OP compounds operating at a low potential. The sensor showed excellent limit of detection, good precision, electrode to electrode reproducibility and stability. The feasibility of application of the sensor for the analysis of real water sample was demonstrated. This method could be extended for the detection of chemical warfare agents and other OP insecticides. The small size, high surface area and other properties of CNT can lead to the development of novel sensors facilitating rapid, on-site monitoring of OP nerve agents with significant implications for homeland security.

Emergent fabrication techniques permit the construction of structures with features on nanometer scale. However, the construction of functional nano-electromechanical systems (NEMS) is hindered by the inability to provide locomotive forces to power NEMS devices. Biomotor molecules are extremely complex machines, and chemists are trying to construct much simpler molecular machines as a logical step toward mimicking the actions of biomotor molecules. The use of biomolecular motors such as enzymes offers an interesting alternative to silicon-based systems. Biomolecular motors such as F1-adenosine triphosphate synthase (F1-ATPase) and myosin are similar in size, and they generate forces compatible with currently producible nano-engineered structures. At Cornell University, for the first time, 7 years ago, hybrid nanodevices were produced by integration of biomolecular motors with nano-engineered systems [230]. The molecular motor was the F1-ATPase from genetically altered Bacillus bacteria, mounted on a 200-nm-high pedestal. It was fueled by adenosine triphosphate (ATP, the so-called energy of cellular life) and it spun nickel nano-propellers at eight revolutions per second, up to a period of two-and-a-half hours.

More recently, it was reported the incrementally staged design, synthesis, characterization and operation of a two-component molecular machine that behaves like a nanoscale elevator [6]. The operation of this device, which was made of a platform-like component interlocked with a trifurcated rig-like component and is only 3.5 nm by 2.5 nm in size, relies on the integration of several structural and functional molecular subunits. This molecular elevator is considerably more complex and better organized than previously reported artificial molecular machines. The energy needed to raise and lower the platform between the two levels on the rig's legs was supplied by an acid-base reaction. The distance traveled by the platform was about 0.7 nm, and it was estimated that the elevator movement from the upper to lower level could generate a force of up to 200 pN. The molecular elevator had a complex structure that is capable of performing well-defined mechanical movements under the actions of external inputs; i.e., according to the authors, it is possible to produce multivalent compounds capable of performing nontrivial mechanical movements and exercising a variety of different functions on external stimulation.

These approaches may enable the creation of a new class of sensors, mechanical force transducers, and actuators. Eventually, nano-biotechnologists are engineering biomolecular motors to run on light energy, with photons instead of ATP. They also plan to add computational and sensing capabilities to the nanodevices, which ideally should be able to self-assemble inside human cells. The hybrid nanodevices produced seem to herald a new generation of ultra-small, robotic, medical devices: *nano-nurses* that move about the body, ministering to its needs, for example; or *smart pharmacies* that detect chemical signals from body cells, calculate the dose and precisely dispense drugs.

5 Cell Adhesion Technologies

The main trends in the BioMEMS research are miniaturization and integration of the biological component with a material surface and the use of microtechniques to improve immobilization and spatial confinement methods. These and other applications requiring the cell/surface interaction account for considerable efforts in development of surface modification and cellular patterning methods, which are very important tools for fundamental studies in biology, especially on single cells, as well as for preparation of chip-based systems in biotechnology.

Substantial progress has been made in surface modification of material structures to control the adhesion of cells. In the current research, the spatially-resolved structuring of surface is basically faced at two different levels: a first, simple *spatial* approach involves featuring the morphology of a surface with topographical properties, as grooves, steps, and trenches. The second *physico-chemical* approach involves the spatially resolved modification of the chemical structure of the surfaces and the related properties. Several methods can be used to tailor the properties of a surface in order to establish the desired interaction between cells and substrates. The methods vary in the degree of control they afford in tailoring the interface. Straightforward approaches that use simple physico-chemical treatments, including plasma oxidation, and/or surface coating through adsorption or chemical grafting of proteins or polymers to the surface.

Surface *bio-functionalization* may provide a way to transform a bio-inert material into a biomimetic or even bioactive material by coupling of protein layers to the surface, or coating the surface with self-assembling peptide scaffolds to lend bioactivity and/or cell attachment 3-D matrix.

Different approaches to physico-chemical functionalization of materials exist. We will address essentially processes of surface modification based on the use of radiation treatments, including essentially low- and medium-energy ion beams [168]. The reason for this is clearly related to the quite good understanding of the beam-induced chemical modifications in polymers [167, 169] and the outstanding capability of focused ion beams of achieving a direct surface patterning [110].

5.1 Experimental Details

In our laboratory, the biology of fundamental interactions between cells and materials is studied in relation to the physico-chemical properties of the biomaterial surface. These properties are determined in terms of chemical composition, wettability and roughness, and modified using both physico-chemical and biological surface functionalization methods. We study cell adhesion in a controlled fashion, using adhesion-supporting and -inhibiting substrata, and analyzing the subsequent cell responses.

The synthetic surfaces used in our investigation included inorganic surfaces as polycrystalline gold (Au), quartz and glass, and model polymers as different as polyhydroxymethylsiloxane (PHMS), polyethyleneterephtalate (PET), polyvinyl chloride (PVC) and polystyrene (PS).

Surface physico-chemical functionalization was performed by irradiation treatments with low-energy ion beams and intermediate fluence regime in order to obtain nanostructuring of surface. In this regime of ion fluence, the local modification of the monomer chemical structure occurs, summing up, until the surface is geometrically saturated, in a process which depends linearly on the fluence. This regime has been therefore indicated as a mild chemistry regime [167]. During this step, a new material of completely different chemical and electronic structure is progressively produced and the related properties, including the optical and electrical ones as well as the surface density of chemical groups, closely reflect the ion-beam induced changes. The produced surface phases are very highly reticulated amorphous phases, which may exhibit signs of nanostructuring, due to the chemical reorganization processes, producing clustering or local densification of the irradiated polymers. A particularly important point is that the interactions of the irradiated surfaces with biological objects as cells, proteins and amino acids are also strongly modified in this regime [168, 208, 210, 214].

In addition, plasma-surface modification has been applied to materials. The unique advantage of plasma processing is that the surface properties and biocompatibility can be enhanced selectively while the bulk attributes of the materials remain unchanged. Existing materials can, thus, be used and need for new classes of materials may be obviated, thereby shortening the time to develop novel and better materials. A plasma, which can be regarded as the fourth state of matter, is composed of highly excited atomic, molecular, ionic, and radical species. It is typically obtained when gases are excited into energetic states by radio frequency (rf), microwave or electrons from a hot filament discharge. Plasma engineering is usually reliable, reproducible, relatively inexpensive and applicable to different sample geometries as well as different materials such as metals, polymers, ceramics, and composite (for a review, [59]). The high-density of ionized and excited species in the plasma can change a variety of surface features of normally inert materials such as polymers, including chemical, tribological, electrical, optical, biological and mechanical properties. When a polymer is exposed to a plasma and if the plasma density and treatment time are proper, many functionalities will be created near the surface and cross-linked polymer chains can be formed. In a typical plasma implantation process, hydrogen is first abstracted from the polymer chains to create radicals at the midpoint of the polymer chains, and the polymer radicals then recombine with simple radicals created by the plasma gas to form oxygen or nitrogen functionalities. Radical species, rather than ion species, that are created in the plasma zone play an important role in the implantation process. Generally, polymers are hydrophobic, and conversion of these polymers from being hydrophobic to hydrophilic usually improves the adhesion strength, biocompatibility, and other pertinent properties. Formation of oxygen functionalities by ion implantation is one of the most useful and effective processes of surface modification. In general, oxygen plasma is used, but plasmas of other compounds consisting of carbon dioxide, carbon monoxide, nitrogen dioxide, and nitric oxide can make the polymer surface hydrophilic as well.

Low energy ion beams are easily adapted to patterning processes, either by using suitable masks, or by using focused ion beams. Also, plasma processes are compatible with masking techniques to enable surface patterning, a process that is commonly used in the microelectronics industry and that, nowadays, finds growing biotechnological applications, as described above. In the last few years, there has been increasing use of microfabrication and nanofabrication technology in the study and uses of cell patterning. This technology has been used for the creation of organized cell cultures for applications such as drug discovery and tissue engineering; moreover, this ability to pattern surface chemistry and topography at subcellular dimensions has enabled new studies in cell biology. On the other hand, the creation of biosensors and cell- based sensors is promoting the development of new technologies for integrating patterned cells with optical and electronic technology. Various forms of coatings and roughening have been used. Conventional photolithography initially was used for chemical and topographical patterning at the micrometer size scale, bringing modern lithographic approaches to these biological problems. The effort to structure surfaces with controlled topography at the nanoscale is a more recent activity. Also, in the last few years, technologies such as contact printing and lift-off approaches have been used for more complex chemical patterning of surfaces. This controlled surface patterning is a component of a new form of biotechnology [69]. In our experiments, the micropatterning relies on the spatially resolved chemical modification of the surfaces, allowing to study, in an exclusive way, the effect of the progressive surface chemistry on the adsorption/organization processes of biomolecules interacting with the irradiated surfaces [209].

More dramatic changes in surface chemistry can be obtained by grafting macromolecules onto the biomaterial. For example, attachment of poly(ethylene oxide) to surfaces has been studied extensively. Although results depend on molecular surface density and chain length, protein adsorption and, subsequently, cell adhesion can be significantly reduced.

Biological surface modification uses understanding of the cell and molecular biology of cellular function and differentiation with the aim to control cell and tissue responses by immobilizing specific biomolecules on surfaces. Since identification of the arginine–glycine–aspartic acid (RGD) motif as mediating adhesion of cells to several plasma and extracellular matrix proteins including fibronectin [115], RGD-containing peptides have been deposited on surfaces to promote cell attachment. Cell surface receptors in the integrin super-family recognize the RGD sequence, and this step is critical to many cellular processes, including cell spreading and integrin-mediated signaling. Another approach to biological surface modification uses whole biomolecules. Whereas depositing small peptides gives the surface a specific characteristic such as cell binding to RGD peptides, immobilizing proteins can provide many functions because of the various domains within the molecule. Intact adhesive proteins, such as fibronectin (FN) and laminin, have been immobilized on biomaterials. Moreover, it is known that upon adsorption to surfaces FN undergoes conformational changes that affect its biological activity [176]. In particular, adsorption of FN to different surfaces alters protein structure and modulates specific integrins binding, cell adhesion, cell spreading and cell migration. The chemical structure and related electronic properties induced by the irradiation treatments used in our experiments could be so effective to support FN correct adsorption and subsequent FN-mediated cell adhesion and patterning. Attachment of growth factors, which can be produced by recombinant DNA techniques, to surfaces has the potential to give biomaterials the ability to induce cell growth, activity, and/or differentiation. Growth factors, such as epidermal growth factor, insulin-like growth factor I, and bone morphogenetic protein 4, have been immobilized on biomaterial surfaces to induce specific cellular responses that cannot be obtained with only adhesion-promoting molecules.

In our studies, biological functionalization of synthetic materials was performed by absorption onto the surfaces of biological molecules, such as RGD, FN or other

ECM proteins, and absorption or binding of phages/peptides selected by phage-display technology to a specific target.

Phage display technology and related applications will be addressed in the last part of the chapter.

Surface modification also can be used to produce surfaces resistant to protein and cell adhesion. A novel physico-chemical treatment will be presented able to obtain anti-bacterial surfaces that in turn enhance mammalian cell proliferation. These surfaces have potential for use in tissue engineering applications.

The surface properties obtained by the above-mentioned functionalization methods were analyzed by X-ray photoelectron spectroscopy (XPS), atomic force microscopy (AFM) and surface free energy (SFE) measurements. The XPS method provides unique information about chemical properties of a surface. It is based upon the photoelectric effect. On irradiation of a sample with a beam of monochromatic X-rays, they penetrate the surface of the specimen and their interaction with the atoms in the specimen causes emission of a core level electron. The energy of this electron is measured and its value provides information about the nature and environment of the atom from which it came. An XPS spectrum represents the photoelectron energy distribution. Because the core electrons of each element have characteristic binding energies, the peaks in the XPS spectra allow identification of all elements, except H and He. With appropriate elemental sensitivity factors, approximate atomic concentrations can be calculated from the relative intensities of the peaks. Additionally, because the electron binding energy is determined by the local chemical environment as well as the type of atom, shifts in the peaks can be used to obtain information about the chemical bonding state of atoms. Thus, from the analysis of XPS scans, we can learn which chemical elements the specimen contains, their ratios and species of each element. XPS has many advantages, including the high information content, the surface localization of the measurement, the speed of analysis, the low damage potential, and the ability to analyze most samples with no specimen preparation.

The AFM is mainly used for topographic surface material characterization. It uses a piezo drive mechanism: the deflection of a tip mounted on a flexible cantilever arm, due to van der Waals and electrostatic repulsion and attraction between an atom at the tip and an atom on the surface, is measured. Atomic-scale measurements of cantilever arm movements can be made by reflecting a laser beam off mirror on the cantilever arm. A one-atom deflection of the cantilever arm can easily be magnified by monitoring the position of the laser reflection on a spatially resolved photosensitive detector. AFM instruments are commonly applied to surface problems using one of two modes, contact mode and tapping mode. In contact mode, the tip is in contact with the surface (excellent topographical imaging can be achieved for rigid specimens in this mode). In tapping mode, the tip is oscillated at a frequency near the resonant frequency of the cantilever. The tip barely grazes the surface; the force interaction of tip and surface can affect the amplitude of oscillation and the oscillating frequency of the tip. The amplitude change translates into topographic spatial information. A major feature of AFM is the ability to acquire three-dimensional images with angstrom- or nanometer-level resolution. Furthermore, imaging can be conducted without staining, coating, or other preparation and under physiological environmental conditions. Moreover, AFM can function well for specimens under water, in air or under vacuum. Striking images of surfaces, biomolecules, and cells can be obtained.

Contact angle analysis involves measuring the angle of contact between a liquid and a surface. The phenomenon of the contact angle can be explained as a balance between the force with which the molecules of the liquid (in a drop sitting on a solid surface) are being attracted to each other (a cohesive force) and the attraction of the liquid molecules for the surface (an adhesive force). An equilibrium is established between these two forces, the energy minimum. The contact angle is an inverse measure of the ability of a particular liquid to wet the surface. If the liquid is water, a smaller contact angle indicates a hydrophilic surface, on which water spreads to a greater extent; a larger contact angle indicates a hydrophobic surface, on which water beads up. Determination of a material's surface energy indicates surface properties better than the mere wettability. Surface energy, defined as the increased free energy per unit area for creating a new surface, is directly proportional to the tendency of molecules to adsorb. Zisman analysis is commonly used to approximate surface energy. Values of contact angles for a series of liquids dropped on the solid are plotted against liquid surface tension; extrapolation of the fitted line to zero contact angle (where complete spreading occurs) gives the critical surface tension value. Experimentally, there are a number of ways to measure the contact angle. In our experiments, the static water contact angles were measured by using the sessile drop method: $2 - \mu l$ liquid drops were applied on at least five different regions of each sample surface and both sides of the two-dimensional projection of the droplets were analyzed by digital image analysis. The water advancing and receding contact angles were measured by the needle-syringe method [182]. The surface free energies, in terms of apolar Lifshitz–van der Waals and polar Lewis acid and basic components, were evaluated using the Lifshitz-van der Waals acid base approach [223], with three test liquids: ultrapure Millipore water, glycerol and tricresyl phosphate.

The actin cytoskeleton is a highly dynamic network composed of actin polymers and a large variety of associated proteins, that mediates a variety of essential biological functions in all eukaryotic cells, including intra- and extracellular movement and structural support. The actin cytoskeleton is a dynamic structure that rapidly changes shape and organization in response to stimuli and cell cycle progression. Orientational distribution of actin filaments within a cell is, therefore, an important determinant of cellular shape and motility. The function of focal adhesions is structural, linking the ECM on the outside to the actin cytoskeleton on the inside, and they are also sites of signal transduction, initiating signaling pathways in response to adhesion. We use immunofluorescence to visualize subcellular distribution of biomolecules of interest. Most commonly, immunofluorescence employs two sets of antibodies: a primary antibody is used against the antigen of interest; a subsequent, secondary, dye-coupled antibody is introduced that recognizes the primary antibody. Typically, this is done by using antibodies made in different species. For example, a researcher might create antibodies in a goat that recognize several antigens, and then employ dye-coupled rabbit antibodies that recognize the goat antibody constant region (denoted rabbit anti-goat). This allows reuse of the difficult-to-make dye-coupled antibodies in multiple experiments. In some cases, it is advantageous to use primary antibodies directly labeled with a fluorophore. This direct labeling decreases the number of steps in the staining procedure and, more importantly, often avoids cross-reactivity and high background problems. As with most fluorescence techniques, a significant problem with immunofluorescence is photobleaching. Loss of activity caused by photobleaching can be controlled by reducing the intensity or time-span of light exposure, by increasing the concentration of fluorophores, or by employing more robust fluorophores that are less prone to bleaching (e.g. Alexa Fluors). In our studies, we perform a very sensitive immuno-cytochemical staining that combines Alexa Fluor-labeled Phalloidin to map the local orientation of actin filaments within cell, a monoclonal antibody to microtubulin that is very specific for the staining of microtubules in cells and the fluorophore DAPI for the fluorescent labeling of nuclei. Immunofluorescent labeled cell cultures were studied using a fluorescence microscope. Fluorescence techniques require special, strong illumination sources, typically mercury or xenon arc lamps; and detection is typically achieved with sensitive imagers, e.g. cooled CCD cameras or photomultiplier tubes. In wild-field epifluorescence microscopy, illumination light from a Hg arc lamp is first passed through a band-pass filter that rejects all light save those wavelengths required to excite the fluorophore (the excitation band pass filter). Next, this excitation light hits a special mirror, the dichroic mirror, located directly behind the objective lens. The purpose of the dichroic mirror is to reflect shorter wavelength excitation light, but allow longer wavelength emission light to pass. The shorter wavelength excitation light is reflected by the dichroic mirror into the objective and delivered to the specimen, where the light is absorbed by the fluorophore and reemitted at a longer wavelength. This light is emitted in all directions; however, a portion of the emitted light is collected again by the objective lens to form the image. This light passes back through the objective and again hits the dichroic mirror. Now, the longer wavelength emitted light is passed by the dichroic mirror and the beam travels onto one last filter (the emission filter, or barrier filter), which rejects all but the emission wavelengths, thus leaving a clean signal that is composed only of light from the excited fluorophore. The three filter elements (excitation filter, dichroic mirror and emission filter) together for a filter set, each specific to the spectral characteristics of one particular fluorophore.

5.2 Some Images

In our studies, we perform immuno-cytochemical staining that combines a mouse monoclonal antibody to microtubulin and a secondary FITC-conjugated anti-mouse antibody for the specific green staining of microtubules in cells, and the fluorophore DAPI for the fluoroescent blue labelling of nuclei (Figs.1,2).

We have, then, performed a multiple fluorescent labeling method by which we can see nuclei in blue, tubulin in red, actin and focal contacts in green and functional mitochondria as bright orange spots (they are labeled by CTC, a tetrazolium salt

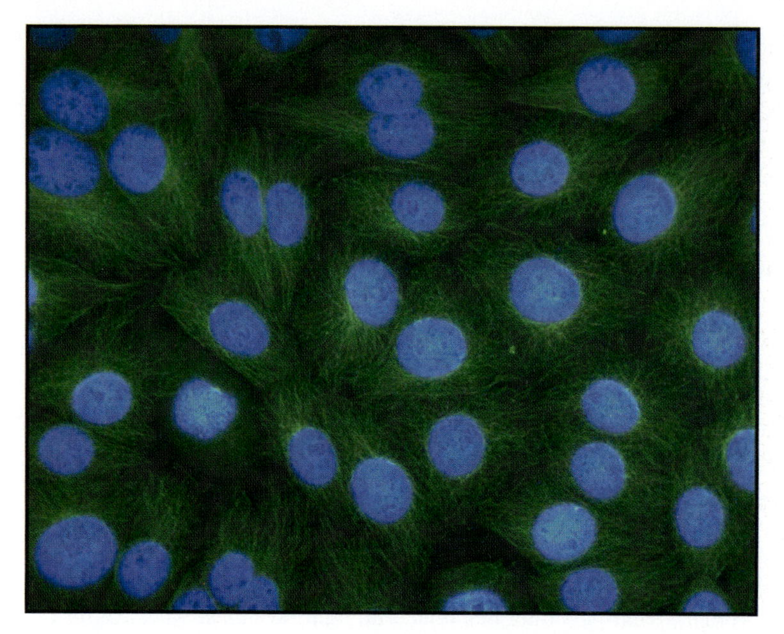

Fig. 1 VERO cells with nuclei and microtubles labelled by immunofluorescence

that becomes orange-fluorescent when reduced by respiratory activity of mitochondria) (Fig. 3). Thus, we have also a functional information about cell adhering on a surface.

With the same technique, we observed normal human dermal fibroblast (NHDF) adhesion on a gold surface and the same surface biofunctionalized with fibronectin [109] (Fig. 4).

Additionally, we obtained high resolution micropatterned surfaces for the creation of organized mammalian cell patterns for applications such as biosensors and, in particular, *single-cell* arrays. We report here biofunctionalization of PHMS

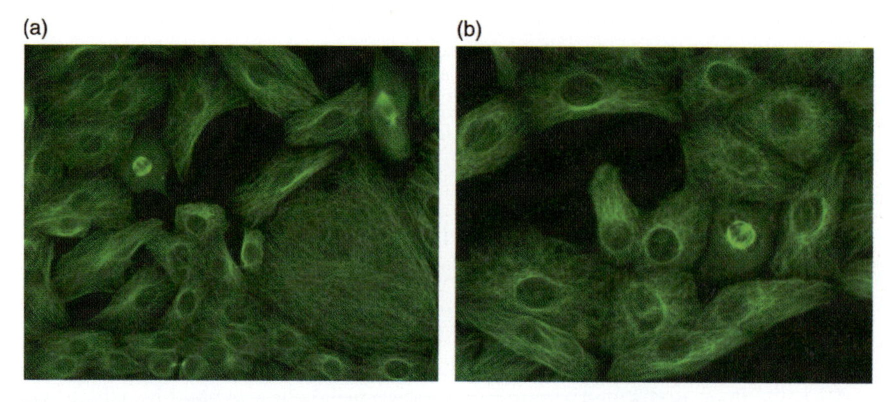

Fig. 2 Mitotic spindles are well visible at 400x (**a**) and 630x (**b**) magnification, thus allowing to assess cell responses to modified surfaces in terms of spreading and proliferation

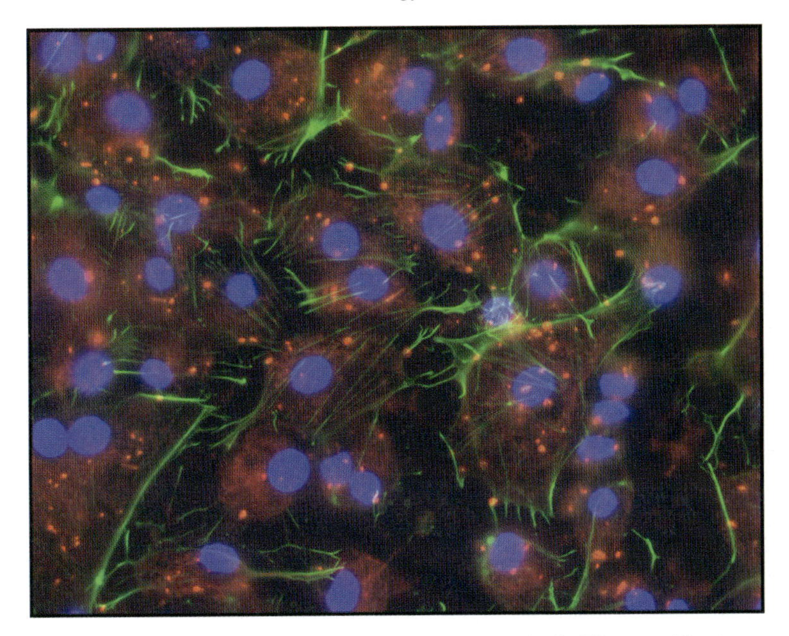

Fig. 3 VERO cells labeled with multiple fluorescent method; 630x magnification

surfaces by controlled ion implantation and fibronectin adsorption aimed to enhance promonocytic cells (U937) adhesion and spatial confinement [212] (Fig. 5).

Guided cell alignment is a fundamental process in development and regeneration. Cell alignment during development appears to be controlled in part by extrinsic guidance cues and also by local influences on cell behaviour resulting from cell-cell interaction. A variety of potential guidance cues are thought to be capable of conveying directional information to cells, likely by operating in concert. In particular, the shape and chemical composition of surfaces can have a bi-directional influence on cell orientation, a phenomenon which was firstly described as *contact guidance* by

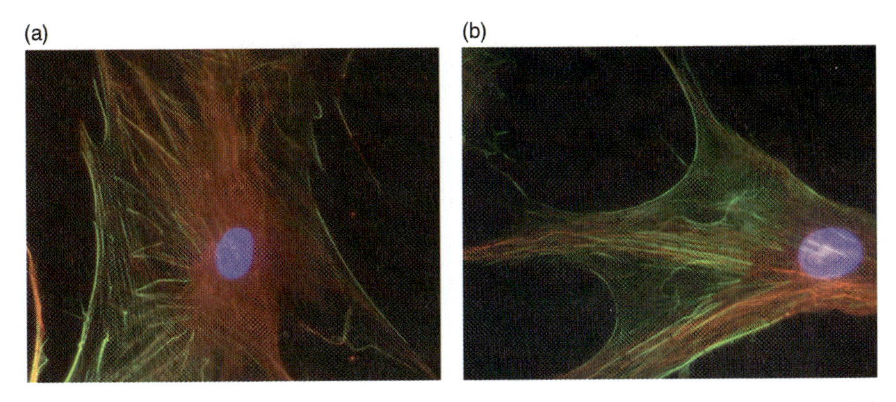

Fig. 4 NHDF cells adhered on gold (**a**) and fibronectin-adsorbed gold (**b**) labeled with multiple fluorescent method

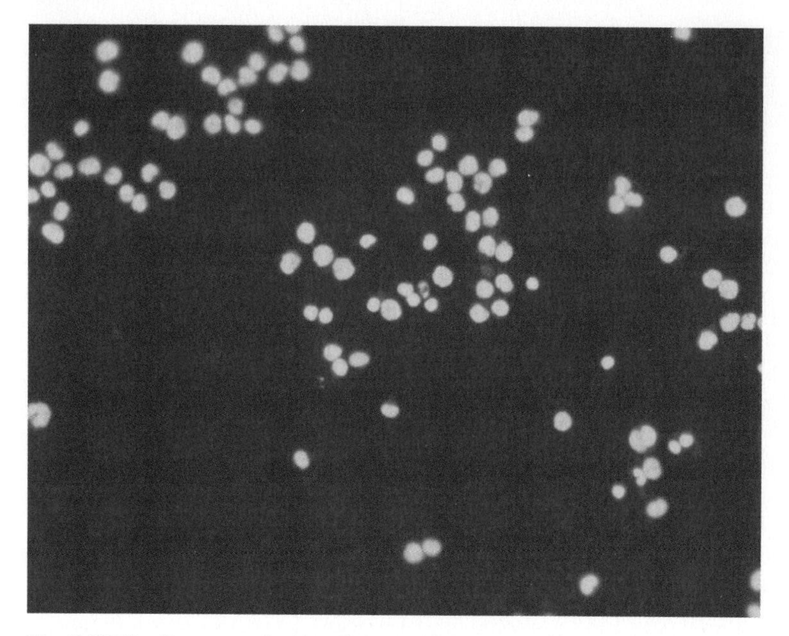

Fig. 5 U937 cells patterned on irradiated and fibronectin-adsorbed PHMS; 200x magnification

Weiss [263–265]. Cell guidance by surface topography [64] and patterned surface chemistry [66] has been studied many times using modified substrates. While this is of interest in developmental biology, identification of synergistic cell guidance properties of substrates could very well improve the performance of prosthetic implants intended to facilitate and promote regeneration or wound repair.

Here, we present the results of a study on NHDF cells seeded onto microfabricated PHMS surfaces [211]. NHDF cells adhered in a selective way onto the patterned surfaces of PHMS, demonstrating the possibility to obtain alignment and controlled positioning of fibroblasts by surface ion irradiation on stripes of given

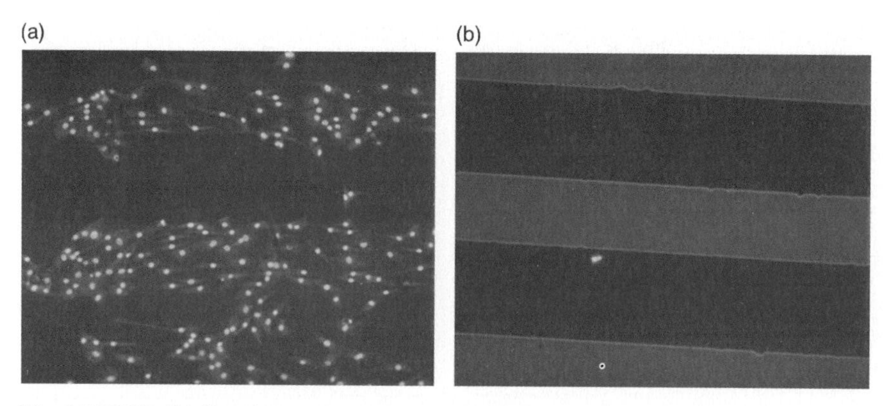

Fig. 6 NHDF cells aligned (**a**) onto micropatterned PHMS (**b**); 200x magnification

dimensions (Fig. 6). Immunocytochemical studies of cytoskeletal components of the NHDF cells were carried out to determine the cellular events underlying the resultant alignment.

In highly aligned cells inside the irradiated stripes, F-actin stress fibers were organized in bundles running parallel to the orientation of the stripes (Fig. 7), whereas they spanned randomly the cytoplasm of cells adhered onto unirradiated surfaces. This coincidence of cytoskeletal maturation and alignment suggested that they are interdependent processes. Co-localization of surface chemistry and stress fibers in aligned cells points to an integrin-based adhesion pathway [249]. Some authors Gingell [112] have speculated that membrane receptors relevant to cytoskeletal formation may accumulate as a response to topographic stimulation. Topography presents powerful cues for cells, and it is becoming clear that cells will react to micrometric [65,79,116,197,221], as well as nanometric [49,75,76,91], scale surface features.

The basis for enhanced guidance observed in our study is likely to be mutual reinforcement of topographic and adhesive effects of ion irradiation. Cells recognize surface features and react to them, resulting in contact guidance. It is known that when a suitable site for adhesion has been detected, focal adhesions and mature actin fibers are formed; tubulin microtubules are then recruited, stabilising the contact [184]. The assembly of FC is itself stimulated by local tyrosine-specific protein phosphorylation, triggered by tension applied to the FC-anchored adhesion complexes by the attached bundles of actin filaments [23]. Our results suggest that the more mature cytoskeleton observed on the irradiated stripes was applying more tension to integrins, resulting in more clustering and larger contacts. Increased integrin clustering may result in enhanced transduction of cell signals to the nucleus, and therefore, fast rates of proliferation and tissue formation. Thus, it appears that both topographic and surface chemical effects of ion irradiation on PHMS are able to alter the morphology and orientation of fibroblasts cells, and hence may be able to control their growth and differentiation. The ability of the substrate to promote the formation of focal contacts and the development of the cell cytoskeleton are important for

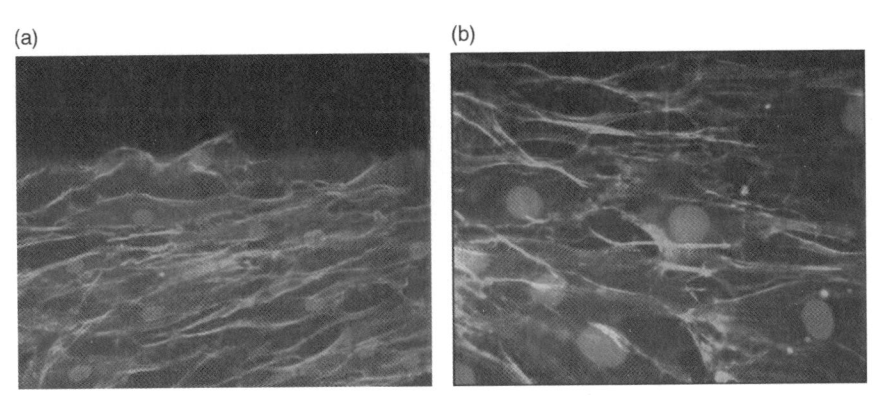

Fig. 7 Alignment of actin fibers is well visible, within cells adhering inside the irradiated stripes, at 400x (**a**) and 1000x (**b**) magnification

the performance of the material. This possibility may have important applications, besides obviously in tissue engineering, also in preparation of cell-based integrated circuits, in a general perspective of bioelectronics devices.

As for bacterial adhesion in relation to properties of functionalized surfaces, we report here images of a novel surface polymer modification able to prevent bacterial colonization while improving controlled cell adhesion [44]. The number of both Gram-positive and Gram-negative bacteria adhered is significantly reduced with respect to the control surfaces (Fig. 8), while VERO cells adhesion and spreading are enhanced on the modified surfaces (Fig. 9). The obtained properties are important issues in development of new biomaterials.

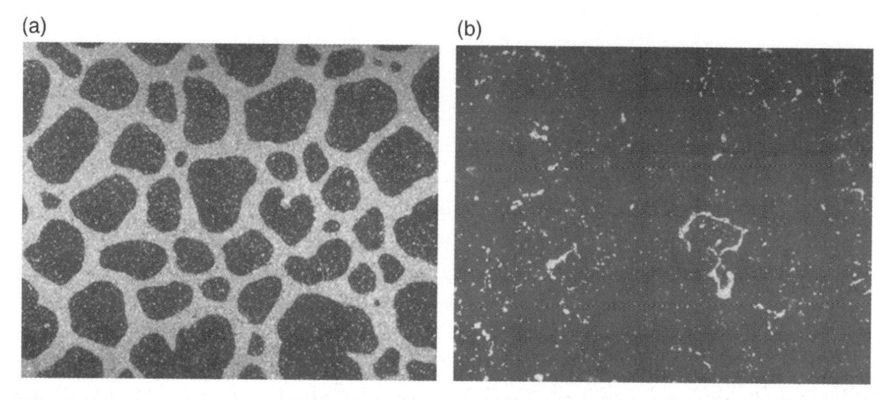

Fig. 8 *S. aureus* onto untreated (**a**) and UV-O_3 treated PHMS (**b**); 1000x magnification

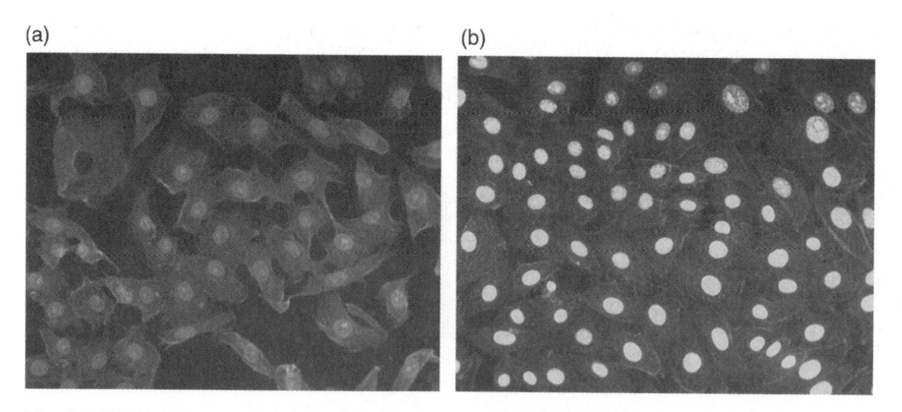

Fig. 9 VERO cells onto untreated (**a**) and UV-O_3 treated PHMS (**b**); 40x magnification

On the other hand, some of the functionalization methods used prompt a massive and unusually fast formation of complex three-dimensional biofilm-like structures [42] (Fig. 10).

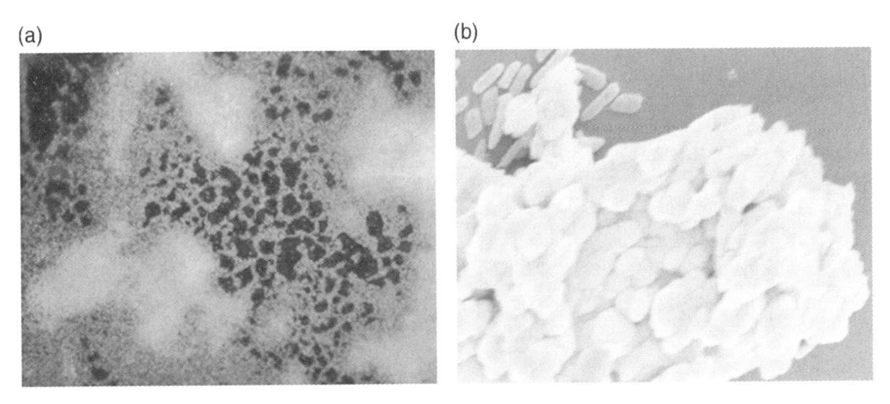

Fig. 10 *P. aeruginosa* biofilm-like structures observed after only 2 hours on irradiated PHMS with epifluorescence microscopy (**a**) and SEM (**b**)

The bacterial adhesion behavior has been studied also onto polymer surfaces nanostructured into bidimensional arrays of nanopores, and a very *active* role of the internal hydrophilic area of nanopores in bacterial response was demonstrated [213](Fig.11).

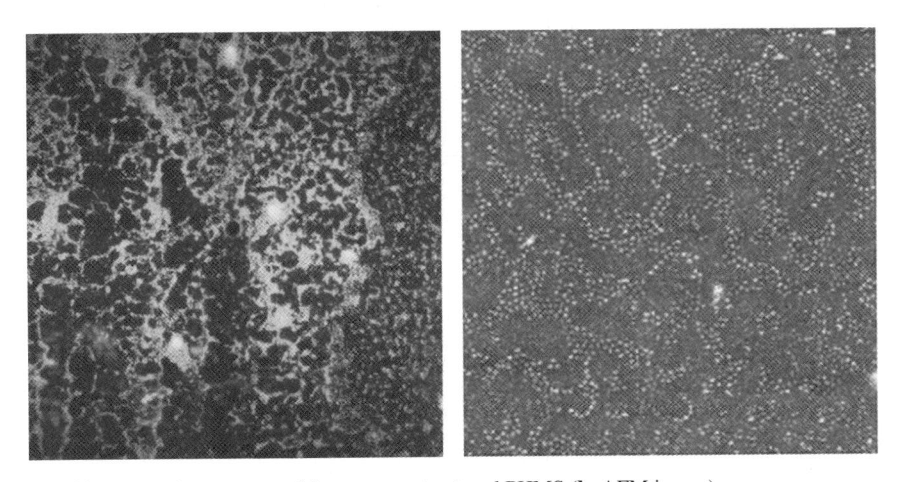

Fig. 11 *P. aeruginosa* pattern (**a**) onto nanostructured PHMS (**b**, AFM image)

At the nanometer scale, a very important aspect must be considered, that is the specificity of nanostructured surfaces in stimulating specific response of large systems as cells, opening new very interesting views on the way the cells can *sense* nanometric surface features.

6 New Advances in Cellular Patterning

One of the crucial features in the fabrication of effective cell-based biosensors is the localized geometric confinement of cells, without losing their viability and sensing capability. The ability to generate even large area patterns of biological materials offers great advantages and opens the door to a wide range of applications in the field of biosensor technology as well as drug discovery.

A number of methods have been reported for patterning biomaterials (mainly cells and proteins). The first class of techniques involves direct transfer of the biomaterial to a substrate using a suitably prepared master stamp. A number of specific methods within this class have been described including microcontact printing, membrane-based patterning, micro-molding in capillaries, and laminar flow patterning. The second class of bio-patterning methods relies on selective chemical or physical modification of the substrate surface to control cell adhesion. This approach has typically relied on linking of the biomaterial through a specific interaction, e.g. a covalent bond or a protein-protein interaction, or manipulation of surface charge, hydrophilicity or topography. Recently, a micro-patterning method of bacterial cells into geometric structural units, defined as corrals, has been reported (Rowan et al. 2002), based on a four-step soft lithographic process.

We found a new approach for cellular patterning using dewetting processes to control microbial adhesion and spatial confinement on modified surfaces. We gave the first demonstration of self-organization prompted by dewetting in microbial distribution on modified surfaces, with coccoid bacteria, rod-shaped bacteria and, very interesting result, the yeast Candida albicans showing similar patterns. We proposed this as a novel potential method for patterning of microbial cells, simpler and more rapid than the usual ones, but reliable and controlled by well defined parameters [45].

An alternative approach for controlled two-dimensional nanoparticle organization on a solid substrate by applying dewetting patterns of charged polymer solutions as a templating system has been recently presented [155]. This method of controlled particle alignment held an attractive feature in its relative simplicity, compared to other methods that require chemical modifications and ligand grafting onto the particles and on the templating substrate. In fact, the template morphology depends on easily controlled parameters, such as the polymer charge density, concentration and dewetting rate, as previously demonstrated by the same authors [154]. Thus, complex morphologies including holes, polygonal networks, bicontinuous structures and elongated structures that are stabilized by viscous forces were produced from dewetting and served as potential templates for nanoparticle organization on a solid substrate.

The particle ordering process was a two-step mechanism: an initial confinement of the nanoparticles in the dewetting structures and self-assembly of the particles within these structures upon further drying by lateral capillary attractions [154], as firstly described for micrometer-size polystyrene latex spheres dewetted on a glass substrate [85]. The dewetting of charged polymer solutions was shown to be a potential method for controlled nanoparticle assembly on a solid substrate. While dewetting morphologies were controlled by polymer solution properties, the templating of the nanoparticle assembly by these polymer films was controlled by the particle size with respect to the thickness of the dewetting film. The easy control of these parameters renders this approach highly attractive as a potential method for directed nanoparticle organization.

In our study, we demonstrated the possibility to control yeast cell organization on polymer surfaces to produce open structures, such as polygonal networks and bicontinuous or elongated structures, using dewetting patterns as templates [45]. We obtained a cellular self-assembly on chemically well-defined solid substrates after controlled dewetting processes. When an aqueous dispersion of particles is deposited on freshly cleaved mica and allowed to dry by evaporation, regions of two-dimensional aggregates are formed, resulting from the self-assembly of particles, a process that has been described by the two-stage mechanism [85]— formation of a nucleus of an ordered phase and convective transport of particles toward the ordered domain. Thus, to form open structures from a drying cell suspension, a control over the nucleation stage of the ordering process is required. The approach adopted in our study involves the control of the nucleation process, by initial cell–substrate interactions, depending on both cell and polymer surface properties. Lateral capillary attractions of the cells confined in the dewetting structures further complete the self-organization process. With this aim, chemically well-defined polymer surfaces were prepared and characterized to be used in cell adhesion and distribution investigation, and the possibility of a selective cellular patterning by means of self-organization phenomena prompted by dewetting processes was demonstrated [45].

Our method offers several advantages over traditional cellular patterning methods. We utilize a method yielding excellent spatial control, to shape and confine our biologically sensitive layer in a pre-casting platform array for potential whole-cell biosensors. We have demonstrated the formation of geometrically well-defined cellular patterns. Single yeast cells were immobilized, forming a high-density fractal array that allows for a scale-free measurement of cell response by a potential transducer. This whole-cell biosensor platform should be very sensitive since each cell acts as an individual, independent sensor, and averaging responses from multiple identical sensors improves the signal to noise ratio. The cellular patterning procedure is simple and the immobilized cells retain their full sensing capabilities.

One of the most intriguing properties of many dynamical systems is their ability to spontaneously generate spatial and spatiotemporal patterns. These have been well studied in fluid dynamics (for example, convection patterns), materials (crystal growth), ecology (formation of herds), to mention only a few examples. Similar types of patterning arise in biological systems, and elucidating the underlying mechanisms responsible for this phenomenon has been the goal of much experimental and theoretical research. There are now a vast number of mathematical and computational models proposed to account for pattern formation in biology, besides a number of different areas.

In order to understand and deeply investigate our patterns of microbial cells with the final aim to control the micropatterning process proposed, we first introduce a model based on self-organization.

Self-organized criticality is an important framework for understanding the emergence of scale-free natural phenomena [34]. It has emerged as an important mechanism for understanding the appearance of scale-free structures in physics and a variety of other disciplines. The idea of self-organized criticality has been applied to problems in a number of fields, including geology [8, 41], cosmology [51] and evolutionary biology [7, 140]. It was demonstrated that dissipative dynamical systems with extended spatial degrees of freedom naturally evolve toward a self-organized critical state that is barely stable, with spatial and temporal power-law scaling behavior [9]. Power law is a unique universal ubiquitarious law governing fractal distributions that is linear on logarithmic scales, where the critical exponent corresponds to the slope of the straight line representing the distribution. Self-organized systems are universal, but the exponents depend on the physical details of each system. In our studies, we used power law to qualitatively assess different dynamic regimes in microbial cell distribution. We demonstrated that a bacterial distribution is fractal when on logarithmic scales it appears to be a power law, i.e. it is fitted in a straight line, and it is characteristic of self-organized criticality; on the contrary, the bacterial distribution is chaotic when it doesn't exhibit a power law behavior, i.e. there are deviations from an exact straight line [43].

The combination of dynamical minimal stability and spatial scaling leads to a *self-similar fractal structure*. Self-similar fractal structures are widespread in nature: the physics of fractals could be that they are the minimally stable states originating from dynamical processes that stop precisely at the critical point. Actions of each component in a fractal distribution are governed by interactions with others, i.e. by a cooperative behavior. To illustrate the basic idea of self-organized criticality in a transport system, Bak et al. [9] introduced the example of a pile of sand. Supposing, to start from scratch, the pile is building by randomly adding sand, a grain at a time. The pile will grow and the slope will increase. Eventually, the slope will reach a critical value; if more sand is added, it will slide off. Alternatively, if the pile is too steep, the pile will collapse until it reaches the critical state, such that it is just barely stable with respect to further perturbations. The critical state is an attractor for the dynamics. As the pile is built up, the characteristic size of the largest avalanche grows, until at the critical point, there are avalanches of all sizes up to the size of the system. The energy is dissipated at all length scales; once the critical point is reached, the system stays there.

Given the complexity of these phenomena, it is important to have a variety of simple models in which the fundamental properties of self-organized criticality can be studied. One of the simplest and most interesting class of models in which self-organized critical behavior can occur are finite state cellular automata. In particular cellular automata (CA), which are a special case of cellular neural networks (CNNs), a well-known paradigm for modelling and emulation of non-linear phenomena in spatially extended systems, provides simple interesting models to study self-organized criticality. CA are discrete dynamical systems constituted by arrays of identical, locally connected *cells* ruled by simple laws [240]. The first CA was introduced by Conway and is the well known Game of Life [22]. Since their origin, CA have been applied to simulating physical and biological systems which are difficult to be modeled by differential equations because of their non-linear and spatially distributed nature [269]. The literature on the use of CA for biological modelling is rapidly growing and it is almost impossible to cite all contributions. As regards bacterial cells, it is worth mentioning the works related to the morphology of bacterial colonies [18, 218] and on the structure of microbial biofilms [135]. We developed a two-dimensional cellular automaton simulation of the microbial patterning process observed that faithfully reproduce our experimental results [35]. In the proposed CA model, the polymer is considered as a grid of compartments (*cells*) in which bacteria may adhere. It is important to distinguish between adhered bacteria and unattached bacteria that can freely move on the polymer surface. Therefore, the status of each *cell* of the automaton represents three different conditions: there is no bacterium in the compartment, there is an adhered bacterium, there is a free bacterium. The CA model is ruled by simple laws that determine how a bacterium reaches the polymer surface and how it adheres to it [35]. More precisely, the law regulating patterning of free bacteria is the following: free bacteria move in random directions and adhere if they encounter other adhered bacteria. This rule accounts for the aggregative behavior of bacteria. Figure shows the result of CA simulation after 100, 200 and 300 steps. Simulation and experimental images perfectly match at every magnification level and also the quantitative comparison, performed by Hausdorff dimension measurement, demonstrates that microbial distribution is not stochastic but ordered in geometrically well-defined patterns. A suitable measure to compare the two pairs of pictures comes from the theory of fractals: the Hausdorff dimension, computed by using the box counting method.

The very important point is that the results of the CA model simulation are a confirmation of the fact that the microbial patterning is the result of a first phase in which microbial cells adhere to the surface and a second, self-organizing process ruled by local laws of interactions. These local laws among the elements of the complex system can account for the formation of the patterns observed.

More recently, to address the understanding of the complex biological phenomena described, we introduced a more complex and detailed computational model, exploiting NetLogo software to compile and simulate an on-purpose developed algorithm [47]. It is an agent-based model where the agents qualitatively mimic the essential features of real bacteria to reproduce the observed behavior: motion and aggregation. We have planned to extract quite abstract principles from biological knowledge and the experiments and apply them to model the phenomenon observed. In this sense, our approach is similar to the so-called *generic modeling* introduced by Tsimring et al. [246], where the biological behavior of bacterial growth is modeled and explained by generic features and basic principles elicited from biological considerations and experimental observations. In particular, several generic models can be grouped into the category of *discrete generic models* [19, 141]. These models share the idea of representing microorganisms by discrete moving entities and of describing the time evolution of nutrients or chemicals with reaction-diffusion equations. We focus on the surface and propose a model based on discrete entities which represent bacteria on it. Each bacterium is essentially a random walker which may adhere on the surface, i.e. stop moving and become fixed at a given position.

The mechanism underlying this process is based on the relative (local) density of the other bacteria. In fact, an element of the model was considered an abstract principle representing the dichotomy between nutrient exchange efficiency and tendency to form bacterial structures. On the one hand, bacteria tend to stay closer and form stable structures that can then evolve into micro-communities and biofilms. On the other hand, bacteria tend to maximize the exchange surface, i.e. to create regions in which the density of bacteria is low and thus rich of nutrients. The access to these regions is favoured by fractal structures as those observed in real experiments. Model simulations matched the experimental results, indicating a threshold behavior with respect to the minimum value of the density of bacteria on the surface needed to observe a self-organized structure. Thus, the idea underlying this rule is that the condition for aggregation is that a sufficient number of bacteria on a given area should exist, otherwise all the bacteria in that area move. We showed that these very simple principles may explain the apparently complex process of microbial patterning.

The cellular patterning procedure we proposed is simple and the immobilized cells retain their full functions and viability [43]. Our data showed that *Staphylococcus aureus* cells were viable in phosphate buffered solution to six hours showing viable cell number less than 10-fold lower than the initial one; then, it halved its number after 24 hours of incubation in buffer. In addition, very similar cellular patterns were obtained when dewetting processes were performed in either buffer or rich culture medium. These results indicated that dewetting process is a very suitable method to obtain patterns of viable microbial cells.

Furthermore, our method allows selective adhesion and patterning of synchronized cells. Intact cells are very attractive candidates for the development of biosensors, because of their highly selective and sensitive receptors, channels, and enzymes; however, one of the major drawbacks of whole cell-biosensors is the heterogeneity of cell population. Individual microbial cells, even those in a *clonal* population, may differ widely from each other in their physiology, biochemistry or behavior. The capability of selectively patterning synchronized microbial cells overcomes this problem. To this aim, we prepared chemically well-defined polymer surfaces allowing the selective adhesion of cells only in the balanced phase of their growth curve, whereas cells from the beginning of stationary phase and thereafter, did not adhere on the modified surface [46].

In addition, by this templating system, large-scale pre-casting microbial arrays can be prepared and stored by lyophilization, holding promise as platforms for BioMEMS devices. We show *S. aureus* patterns obtained on quartz maintained 90% cell viability, assayed by LIVE/DEAD kit, after lyophilization [46].

In conclusion, we proposed this new technology as a useful tool for preparation of microbial arrays that hold promise as platforms for a number of biotechnological applications. For example, *C. albicans* cells on plasma-treated PET form, after dewetting, a typical hexagonal array very similar to that obtained by artificial micropatterning [45].

The control of dewetting behaviour to produce cellular patterns offers a potential templating system for preparation of large-scale microbial arrays that hold promise as platforms for BioMEMS devices, as diagnostic chips for detection of chemical pollutants and/or toxins in the environment (biosensing), and HTS applied to drug discovery. In fact, it should be possible, for example, to prepare large libraries of mutant microorganisms, pattern them into well-defined fractal arrays, and then individually examine the arrays when they are dosed with small molecules such as drug candidates or environmental analytes. Microfluidic systems are ideally suited for implementing such assays on bioarrays such as these.

6.1 Perspectives of Microbial Arrays

Microbial arrays obtained with our proposed novel technology of micropatterning are characterized by the highest cell density on the least surface area, and this very important property makes them attractive candidates as platforms not only for whole-cell biosensors and diagnostic chips, but also for other topical applications such as microbial fuel cells and microbial filters.

6.1.1 Microbial Fuel Cells

A microbial fuel cell (MFC) is a device that directly converts chemical energy to electricity by the catalytic reaction of microorganisms on the anode side. The driving force is the redox reaction of a carbohydrate substrate, such as sugars and alcohols, using a microorganism as catalyst, therefore noble metal is not needed, and working conditions are mild— ambient temperature, normal pressure and neutral pH. The existing MFCs have shown good stability for long operating periods, the runs have lasted for over one month. Moreover, the reactions are self-regulating, and self-discharge is negligent. Reaction does not require heat exchange or cooling down.

There are two types of microbial fuel cells. One involves the utilization of electroactive metabolites, e.g., hydrogen, converted by microbial metabolism from substrate, and another involves the utilization of mediators as electron transporters from a certain metabolic pathway of the microorganism to electrodes.

A typical MFC consists of anode and cathode compartments separated by a cation-exchange membrane (CEM). In the anode compartment (with a gas sparger to remove air), fuel is oxidized by microorganisms on the anode surface generating electrons and protons. Electrons are transferred through an external circuit to the cathode and create current, and the protons through the membrane to sustain the current. Typically, electrons and protons are consumed in the cathode compartment reducing oxygen to water by a catalyst such as platinum. The electron transfer can occur either via membrane-associated components, soluble electron shuttles or nanowires.

After using baker's yeast as microorganism in the MFC a few years ago, bacteria have been used. One of the main reasons to use bacteria is that they could survive in a higher pH solution; this offers a lower anodic potential, which makes a higher potential difference in the terminals of the fuel cell. The electrochemical reaction and the reduction of terminal electron acceptors typically occur inside the bacterial cells and most of the microbial cells are electrochemically inactive. In their model, Zhang and Halme [279] assumed that electron transfer to the electrode surface was carried out by an electron-transfer mediator. The electron transfer from microbial cells to the electrode, in fact, can be facilitated by mediators such as potassium ferric cyanide, thionine, methyl viologen (methyl blue), humic acid, neutral red, etc. [84, 158]. Most of the mediators available are expensive and toxic. Recent discoveries, however, revealed that some dissimilatory metal-reducing bacteria such as *Geobacter* and *Shewanella* species can directly transfer electrons out of their cell onto solid electrode surfaces [25]. Biochemical and genetic characterizations indicated that outer-membrane cytochromes can be involved in exogenous electron transfer [165, 177, 180]. Among the electrochemically active bacteria are *Shewanella putrefaciens* [142], *Aeromonas hydrophila* [73] and others. Also, some bacteria produce and use soluble electron shuttles that eliminate the need for direct contact between the cell and electron acceptor [248]. For example, phenazine production by a strain of *Pseudomonas aeruginosa* stimulated electron transfer for several bacterial strains [196]. This extracellular electron transfer facilitates the use of these bacteria as catalysts in electrochemical cells which oxidize a variety of electron donors and to capture the electrical energy produced. On the other hand, the recent discovery of nanowires introduces a whole new dimension to the study of extracellular electron transfer. These conductive, pilus-like structures, identified so far in *Geobacter sulfurreducens* PCA [199], *Shewanella oneidensis* MR-1 [114], a phototrophic cyanobacterium *Synechocystis* PCC6803 [114], and the thermophilic fermenter *Pelotomaculum thermopropionicum* [114], appear to be directly involved in extracellular electron transfer. Disruption of a pili gene in *G. sulfurreducens* eliminated the bacterium's ability to reduce insoluble electron acceptors [199]. Deletion of the genes associated with two *c*-type cytochromes (MtrC and OmcA) in *S. oneidensis* resulted in poorly conductive nanowires, loss of electrochemical activity, and loss of the ability to reduce insoluble electron acceptors [114]. These nanowire structures allow the direct reduction of a distant electron acceptor. This removes the need for soluble mediators that would be lost in a continuous-flow MFC and may allow for direct interspecies electron transfers.

A photosynthetic electrochemical cell (PEC) uses a photosynthetic microorganism (or its subcellular components) to generate electricity. *Synechococcus* sp [273] and *Anabaena variabilis* [237] have been studied. It was understood that PECs generate electricity from both photosynthesis and catabolism of endogenous carbohydrates in the light and from catabolism alone in the dark. Provided with light, the photosynthetic bacterial cells carry on the reactions of photosynthesis, converting CO_2 and H_2O into O_2 and carbohydrates (e.g. glucose). During photosynthesis, electrons are being shuttled in the diffusional electron carriers NADPH or along a series of thylakoid-membrane-bound enzyme complexes of the electron transport chain. These electrons (and protons) are siphoned from their normal photosynthetic duties either from NADPH or the transport chain by redox electron mediator molecules that have diffused into the bacterial cells. Then, these reduced (electron- and proton-carrying) mediators make their way back by diffusion out of the bacterial cells, through the buffer solution, and eventually donate the electrons to the anode. Just as in the MFC, the electrons then travel through an external load into the cathode chamber, where they reduce the oxidant ferricyanide (Fe(III)). The protons cross the PEM from the anode into the cathode, where they combine with the reduced oxidant (Fe(II)) or with O_2 and electrons from the reduced oxidant to release H_2O.

It is clear that more work is needed to develop a comprehensive model of MFCs in order to improve our understanding of MFCs and to optimize their design and operation.

Microbial fuel cells have a number of potential uses. The advantages to using a MFC as opposed to a normal battery is that it uses a renewable form of energy and would not need to be recharged like a standard battery would. Further, they could also be built very small [54] and they operate well in mild conditions, 20°C to 40°C and also at pH of around 7 [36]. The first and most obvious use is harvesting the electricity produced for a power source. Virtually any organic material could be used to *feed* the fuel cell, and a MCF is more efficient than standard combustion engines that are limited by the Carnot Cycle. In theory, a MFC is capable of energy efficiency far beyond 50% [278]. The most immediate and useful applications for MFCs are in wastewater treatment [195] and as power sources for environmental sensors [111], but opportunities for other applications exist, most recently in biomedicine, as power sources for pacemakers and microdevices, using glucose or other substrates from the blood stream [54]. With modifications, MFC technologies could find applications ranging from H_2 production [159] to renewable energy production from waste biomass [283]. Moreover, the basic system could be modified for environmental bioremediation: power can be put into the system to drive desired reactions to remove or degrade chemicals. Bacteria are not only able to donate electrons but can also accept electrons from the cathode. By poising the electrodes at –500 mV, Gregory and Lovley [118] were able to precipitate uranium directly onto a cathode because of bacterial reduction. Nitrate can also be converted to nitrite when electrodes are used as electron donors [117]. Electrolytic cultivation has been used to extend the growth rates of suspensions of iron-oxidizing bacteria in the laboratory [171].

Different shapes of the anode and the cathode are studied. It seems that the thickness of the anode compartment should be less than 2 cm and larger membrane area is needed for a high energy output. On the other hand, the size of the biological fuel cell is important factor for the efficient energy conversion from chemical to electrical energy. The smaller size of the fuel cell improves the efficiency. Miniaturized, portable power sources are important components for the realization of complete microsystems. Micromachined microbial fuel cells (μMFCs and μPECs) were first reported by Chiao et al. [57] and then optimized in their operation by the same researches [58]. A bulk micromachining process was used to fabricate the fuel cells, and the prototype had an active proton exchange membrane area of 1 cm^2. Two different microorganisms were used as biocatalysts in the anode: *Saccharomyces cerevisiae* (baker's yeast) to catalyze glucose, and *Phylum Cyanophyta* (blue-green algae) to produce electrons by a photosynthetic reaction under light. In the dark, the μPEC continued to generate power using the glucose produced under light. In the cathode, potassium ferricyanide was used to accept electrons and electric power was produced by the overall redox reactions. The μMFCs and μPECs may find applications in powering electronic and MEMS devices. Although the power density reported is low and cannot find practical applications, with the advances in biochemistry and microbial fuel cell field, researchers can one day optimize microbial chemistry that could be used in μMFCs and μPECs.

Challenges remain for MEMS researchers to build small-scale anode and cathode chambers that are suitable for specific microorganism/mediator combination and to develop a scalable or modular MFC-based technology that can provide, in a cost-effective manner, the large microbial surface areas needed. Our proposed novel micropatterning technology allows us to obtain microbial arrays with properties that make them very attractive to meet these requirements.

6.1.2 Microbial Biofilters

A biofilter is one of several air pollution control technologies that use microorganisms to treat odorous air. Microorganisms in biofilters perform the removal and oxidation of compounds from contaminated air. Smelly air emissions generally contain low concentrations of hydrogen sulfide, mercaptons and other reduced sulfur compounds, which the microorganisms use as food for energy and nutrients. The by-products are primarily water, carbon dioxide, mineral salts, some organic compounds and more microorganisms.

Patent applications for odor control using biofiltration have been filed since the 1950s for soil filters and large biological trickling filter plants. Biofilters have been designed primarily for odor control at wastewater treatment plants, rendering plants and composting operations. During the 1990s, biofilters were also used to remove airborne contaminants including aliphatic and aromatic hydrocarbons, alcohols, aldehydes, organic acids, acrylate, carbolic acids, amines and ammonia. They are becoming more popular in the treatment of volatile organic compounds, as an innovative method to treat toxic air emissions from commercial processes. These substances are not just smelly, they are dangerous as well. A well-managed biofilter can reduce odor emissions by 85%, hydrogen sulfide by 90% and ammonia by around 60%. Emission reductions can vary widely from 20% to nearly 100%. Biofilter media moisture content and residence time (the time required for the air to pass through the biofilter media) are key factors that affect effectiveness. Based on research and reports on industrial biofilters made using wood chips and compost, they are expected to last 5 to 10 years.

Previously, biological waste gas purification has been reported for several air pollutants [88, 121] including ethylene (C_2H_4) [83, 252]. A biofilter which eliminated C_2H_4 from the high parts-per-million range to levels near the limit for plant hormonal activity (0.01 to 0.1 ppm) was developed [94]. Isolated ethylene-oxidizing bacteria were immobilized on peat-soil in a biofilter and subjected to an atmospheric gas flow with C_2H_4. Ethylene was removed to 0.017 to 0.020 ppm. Reduction of C_2H_4 to such extremely low levels is attractive at industrial point sources and is an important prerequisite for the development of a biofilter for use in horticultural storage facilities. Other characteristics of the biofilter which were favorable for such use included the operational stability extended for more than 75 days, the removal adapted at 10°C, and storage of the inoculated peat-soil for 2 weeks at 20°C. A recent study demonstrates that biotrickling filters can replace chemical scrubbers and be a safer, more economical technique for odor control [107].

The purpose and underlying concept of in-situ bioremediation (ISB) is to attenuate hazardous compounds in the soil by bio-transforming these substances into innocuous forms. The first commercial ISB system was implemented in 1972 at a pipeline failure site in Pennsylvania. ISB system involves the addition of nutrients and suitable electron acceptors to the contaminated soil to promote the breakdown of the contaminants by microorganisms in place. ISB does not include any treatments that require excavation of contaminated soil or pumping of groundwater to a treatment system. The in-situ microbial filter (ISMF) represents a specific application of ISB. This application involves placing sand mixed with non-indigenous microorganisms into a trench in the subsurface ahead of the contaminant plumes. The contaminants in the groundwater are metabolized by the microorganisms as the groundwater flows through the trench. This remediation method can reduce or eliminate groundwater contamination, thus reducing the need for extensive monitoring and treatment requirements. ISMFs have the following advantages: removal of contaminated soil is only limited to the trench excavation; contaminants in the subsurface are degraded in-situ instead of transferring them to another medium; a continuous input of energy for pumping groundwater is not required, thus the system will not be prone to failure due to mechanical breakdown or power outage; the filters will continue to operate with only minor inputs of energy to supply the required oxygen and nutrients, and maintenance is limited; and, since water is not brought to the ground surface for treatment, technical and regulatory problems related to discharge of treated water are avoided and scarce groundwater resources are not wasted. ISMFs are a relatively inexpensive treatment method for the above reasons. Also, using an ISMF will often allow remediation to proceed without interrupting the normal site activities, and there are no additional costs due to temporary storage or transportation of contaminated soil. In addition, the sandy soil biofilter material will not need removal once remediation is complete. One study focused on the use of ISMFs for the remediation of naphthalene-contaminated groundwater [261]. However, till now biofilters are used mainly for odor control or air pollutants decontamination for problems of blocking and flux control in ISMFs. The ability to obtain ordered microbial arrays with a fractal geometry by our proposed patterning technology could overcome these problems and allow biofilter use also for liquid decontamination.

7 Bio-functionalization by Phage-Displayed Peptides

Finally, perspectives are presented of surface bio-functionalization by phage-displayed peptides, which can act as highly specific and selective probes in bioaffinity sensors, can be used in development of nanomaterials and cantilever-based nanodevices for biosensing, and can mimic ligands of cell receptors involved in signaling that affect the cellular fate.

7.1 Phage Display Technology

We used phage display technology, a powerful tool for selection of short random peptides with high affinity to target structures of interest. Phage display technology involves the expression of random peptides on the surface of a filamentous bacteriophage, M13, displayed as a fusion with one of the viral structural protein, either the terminus pIII protein (5 copies) or the major coat protein pVIII (2700 copies). By cloning large numbers of DNA sequences into the phage, display libraries are produced with a repertoire of many billions of unique displayed proteins (98, 161). Recombinant peptides specifically binding a target of interest can be selected from random peptidic libraries (usually from 9- to 15-mer), by a process of affinity selection known as *biopanning*. Briefly, a simple method for biopanning involves incubating the library with the target (either a single receptor tethered to a solid support or intact cells for selection of tissue and cell targeting proteins) to allow phage displaying a complementary protein to the target to bind. Non-binding phages are then washed away and those that are bound — usually a tiny minority —are eluted in a solution that loosens target-peptide bonds. The eluted phages are still infective and are propagated simply by infecting fresh bacterial host cells, yielding an *amplified* eluate that can serve as input to another round of affinity selection. Successive rounds of biopanning enrich the pool of phage, with clones that specifically bind the target. Phage clones from the final eluate (typically after 2–4 rounds of selection) are propagated and characterized individually. The amino acid sequences of the peptides responsible for binding the target receptor are determined simply by ascertaining the corresponding DNA coding sequence in the phage genome.

The proteins displayed range from short amino acid sequences to antibody fragments, enzymes, cDNA and hormones. Phage display is an exponentially growing research area, and numerous reviews covering different aspects of it have been published [67, 99, 229].

7.2 Phage-Displayed Peptides as Diagnostic Probes

Advanced bioselective sensors may meet the requests for isolation, concentration of the agents and their immediate real-time detection. The majority of rapid detection biosensors described in the literature has utilized antibodies as bioreceptors [160]. However, while sensitive and selective, antibodies have numerous disadvantages for use as diagnostic biodetectors in food products, including high cost of production, low availability, great susceptibility to environmental conditions [226] and the need for laborious immobilization methods to sensor substrates [190]. An effective alternative to antibodies may be short peptides affinity-selected from random phage-displayed peptide libraries for specific, selective binding to biological targets [231]. Filamentous bacteriophages can display on their surface foreign peptides, expressed by foreign DNA introduced in the genome through recombinant modification. In this way, recombinant phage clones may recognize and bind specific targets, such as cell surface receptors. Thus, they can act as antibody surrogates, possessing distinct advantages including durability, stability, standardization and low-cost production, while achieving equivalent specificity and sensitivity [190, 191]. In addition, the short protein structure (the outer coat protein structure of filamentous phage) appears to be highly amenable to simple immobilization through physical adsorption directly to the sensor surface, thus providing another engineering advantage while maintaining biological functionality [48, 186]. Numerous phage applications have been proposed, including the detection of small molecule [203], receptors [10], and whole-cell epitopes [48,78,186,189,231,235,277]. In particular, this technology represents a powerful tool for the selection of peptides binding to specific motifs on whole cells since it is a non-targeted strategy, which also enables the identification of surface structures that may not have been considered as targets or have not yet been identified [24].

In our studies, we used a whole-cell phage display approach to isolate peptides specifically binding to surface of bacterial cells [48]. We demonstrated, along with other previously published data, that phage-displayed peptides show promise as probes for biosensor applications. In fact, these phage probes could be used to build micro-biosensor systems in which biological sensing element is the selected phage-displayed peptide.

Peptides selected by phage-display may find application as biosorbent and diagnostic probe for monitoring bacterial cells by various devices in which antibodies have been used to date. The potential advantages of phage-displayed peptides as replacements for antibodies commonly used in immunoassays include the simplicity of manipulation of the phage libraries, their great variability, high binding affinity, low steric hindrance, great stability and the possibility of selecting probes to targets of different nature, also to small molecules or toxic compounds or immunosuppressants against which it is difficult to raise natural antibodies. For these properties, they may be exploited for development of bioaffinity sensors, whose essential elements are probes that specifically recognize and selectively bind target structures and, as parts of the analytical platform, generate a measurable signal. For example, they may be used for separation and purification of bacteria prior to their identification with polymerase chain reaction, immunoassays, flow cytometry, or other methods. Furthermore, they may find application as biorecognition elements of real-time biosensor devices.

Recombinant peptides selected by phage-display selectively recognize and specifically bind complex structures such as bacterial cells. Thus, they can be used to develop rapid diagnostic arrays. In fact, traditional diagnostic systems usually involve a multi-step detection method with the use of labeled secondary antibody, whereas phage-displayed detection microsystems could be considered one-step, simultaneously bind and identify the target microorganism, with no need of further characterization steps.

In addition, phage-displayed peptides can functionalize surface with less steric hindrance than antibodies, thus allowing a higher binding avidity for the target per surface unit. In fact, on the same surface unit, a greater number of peptides and with a more correct orientation can be patterned in comparison to antibodies. Furthermore, the nature of the bioreceptor peptide holds potential utilization for development against any bacterium, virus or toxin to which a corresponding phage could be affinity-selected for. Therefore, different peptides could be isolated specifically binding to isolated proteins, enzyme or inorganic material, as well as to different microbial species, thus with the same microsystem different enzymes, toxins and pathogens might be detected, by performing parallel several different assays in real-time, within the same miniaturized substrate, in a single run. This could ultimately translate to a much lower cost per test. Much of the promise of these microarrays relies on their small dimensions, which reduce sample and reagent requirements and reaction times, while increasing the amount of data available from a single assay. Through the use of different labels in parallel, such as different specific peptides, multiple tests could be simultaneously performed on the same microarray in a single step, so that standardizing data from multiple separate experiments is unnecessary and truly meaningful comparisons can be made. The development of highly sensitive and accurate field-usable devices for detection of multiple biological agents could have a number of applications in biomedical field as well as in monitoring of agrofood pathogens and detection of biological warfare agents.

7.3 Phage-Displayed Peptides Binding Materials

Another application of particular value for phage display technology would be methods that could be applied to functionalize materials with interesting electronic or optical properties. Although natural evolution has not selected for interactions between biomolecules and such materials, phage-display libraries can be successfully used to identify, develop and amplify binding between organic peptide sequences and inorganic semiconductor substrates. Peptides with limited selectivity for binding to metal surfaces and metal oxide surfaces have been successfully selected [31, 32]; other researchers have used phage display to select peptides against synthetic polymers such as polystyrene [1] and yohimbine-imprinted methacrylate polymer for molecular-imprinted receptors [21]. This approach was then extended and it was shown that combinatorial phage-display libraries can be used to evolve peptides that bind to a range of semiconductor surfaces with high specificity, depending on the crystallographic orientation and composition of the materials used [266]. Phage-display libraries, based on a combinatorial library of random peptides — each containing 12 amino acids — fused to the pIII coat protein of M13, provided 10^9 different peptides that were reacted with crystalline semiconductor structures. Five copies of the pIII coat protein are located on one end of the phage particle, accounting for 10–16 nm of the particle. The phage-display approach provided a physical linkage between the peptide-substrate interaction and the DNA that encodes that interaction. The experiments utilized different single-crystal semiconductors for a systematic evaluation of the peptide-substrate interactions. Crystal-specific phages were isolated and their DNA sequenced. Peptide binding selective for the crystal composition (for example, binding to GaAs but not to Si) and crystalline face (for example, binding to GaAs (100), but not to GaAs (111)B) was demonstrated. In addition, the preferential attachment of phage to a zinc-blended surface in close proximity to a surface of differing chemical and structural composition was reported, demonstrating the high degree of binding specificity for chemical composition.

Subsequently, phage display has been used again in selecting unique peptides against inorganic semiconductor materials [101,205]. Reviews [206,207] have highlighted the application of phage display in selecting peptides to functionalize biomaterials such as titanium. More recently, a unique strategy for surface functionalization of an electrically conductive polymer, chlorine-doped polypyrrole (PPyCl), which has been widely researched for various electronic and biomedical applications, has been developed [204]. A M13 bacteriophage library was used to screen 10^9 different 12-mer peptide inserts against PpyCl, a binding phage was isolated, and the stability and specificity, strength and mechanism of its binding to PPyCl were assessed. In these studies, phage display was used to select peptides that specifically bound to an existing biomaterial, PPy, and were subsequently used to modify the surface of PPy. PPy is a conductive synthetic polymer that has numerous applications in fields such as drug delivery [146] and nerve regeneration [219, 251], and has been used in biosensors and coatings for neural probes [70, 256]. Different dopant ions such as chloride, perchlorate, iodine and poly (styrene sulphonate) can be used during electrochemical synthesis to provide the material with varying properties (for example, conductivity, film thickness and surface topography). PPyCl does not contain a functional group for biomolecule immobilization, making it a suitable model polymer for functionalization using a peptide selected with phage display. Further, the specific peptide selectively binding PPyCl was joined to a cell adhesive sequence and used to promote cell attachment on PPyCl, to serve as a bifunctional linker. The use of the selected peptide for PPyCl by phage display can be extended to encompass a variety of therapies and devices such as PPy-based drug delivery vehicles [146], nerve guidance channel conduits [219,251], and coatings for neural probes [71]. Furthermore, this strategy for surface functionalization can be extended to immobilize a variety of molecules to PPyCl for numerous other applications. In addition, phage display can be applied to other existing polymers (including those that are already approved and/or those polymers that lack functional chemical groups for coupling reactions like PPyCl) to develop bioactive hybrid-materials without altering their bulk properties.

Selection of peptides using phage display thus represents a simple and versatile alternative to methods based on electrostatic and hydrophobic interactions between two moieties to achieve functionalization of surfaces. It is theoretically possible to design bivalent recombinant phage with two-component recognition: such phages have the potential to bind to specific locations on a semiconductor structure by peptides displayed on pIII protein and simultaneously to specific target (molecules or cells to be captured) by peptides displayed on pVIII coat protein.

Given that filamentous phage are resistant to harsh conditions such as high salt concentration, acidic pH, chaotropic agents, and prolonged storage, they are suitable candidate building blocks to meet the challenges of bottom-up nanofabrication. Moreover, the pIII minor capsid protein of the phage can be easily engineered genetically to display ligand peptides that will bind to and modify the behavior of target cells in selected tissues. Thus, the tactic of integrating phage display technology with tailored nanoparticle assembly processes offers opportunities for reaching specific nanoengineering and biomedical goals [5, 113, 149, 153, 242]. Recently, an approach for fabrication of spontaneous, biologically active molecular networks consisting of phage directly assembled with gold (Au) nanoparticles has been reported [233]. In this work, it was shown that such networks are biocompatible and preserve the cell-targeting and internalization attributes mediated by a displayed peptide and that spontaneous organization (without genetic manipulation of the pVIII major capsid protein), and optical properties can be manipulated by changing assembly conditions. By taking advantage of Au optical properties, Au–phage networks were generated that, in addition to targeting cells, could function as signal reporters for fluorescence and dark-field microscopy and near-infrared (NIR) surface-enhanced Raman scattering (SERS) spectroscopy. Notably, this strategy maintains the low-cost, high-yield production of complex polymer units (phage) in host bacteria and bypasses many of the challenges in developing cell-peptide detection tools such as complex synthesis and coupling chemistry, poor solubility of peptides, the presence of organic solvents, and weak detection signals. These networks can effectively integrate the unique signal reporting properties of Au nanoparticles while preserving the biological properties of phage. Together, the physical and biological features within these targeted networks offer convenient multifunctional integration within a single entity with potential for nanotechnology-based biomedical applications.

A research project we are working is based on the possibility of using M13 phage for production of new optoelectronic nanomaterials due to formation of supramolecular aggregates of porphyrins through the controlled interaction with recombinant M13 phages. AFM analysis of phage on gold substrates, in association with porphyrin molecules, shows the formation of ordered supramolecular aggregates controlled by phage. The second part of the project deals with the possibility of producing nanodevices for biosensing in biomedicine based on cantilevers functionalized with recombinant M13 phages specifically binding biomolecules, such as antibodies. The extent and rate of cantilever deflection, as a result of the specific binding between phages and target biomolecules, are determined by AFM techniques (laser source and photodetector). Sensibility of the method allows to discriminate also among the different classes of antibodies that, for their different molecular weight, cause a different cantilever deflection degree.

7.4 Phage Functionalization of Carbon Nanotubes

New perspectives in phage-based nanomaterials and nanodevices involve the use of carbon nanotubes (CNTs). Among the anticipated applications of CNTs is their use as components in biological devices. The possibility of the promotion of electron transfer reactions at a lower potential due to their structure dependent metallic character and their high surface area provide ground for unique biochemical sensing systems.

Until now, most of the research undertaken in this direction has been focused on attaching biological molecules onto these nanomaterials. It has been shown that small proteins can be entrapped into the inner channel of opened nanotubes by simple adsorption [81, 244, 245]. Attachment of small proteins on the outer surface of carbon nanotubes has also been achieved, either by hydrophobic [12] and electrostatic interactions [52], via covalent bonding [127] or by functionalization of the nanotube sides by polymer coating [225]. Bridging nanotubes with biological systems, however, is a relatively unexplored area, with the exception of a few reports on nanotube probe tips for biological imaging [270], nonspecific binding (NSB) of proteins [12, 97, 225], functionalization chemistry for bio-immobilization on nanotube sidewalls [52], and one study on biocompatibility [172]. More recently, high specific anti-*Salmonella* and anti-*Staphylococcus aureus* antibodies immobilization on hydrophobic and hydrophilic nanodiamonds and CNTs coated silicon substrates was reported [126]. The efficacy of both antibodies immobilization and bacterial binding on air plasma treated nanodiamond is better than those of the hydrogen plasma treated, because the heat treatment in air causes the oxidation of the CNTs that become more hydrophilic, terminating with O and OH groups.

Chen et al. [53] systematically explored how nanotubes interact with and respond to various proteins in solution, how chemical functionalization could be used to tailor these interactions, and how the resulting understanding enables highly selective nanotube sensors for the electronic detection of proteins. NSB on nanotubes, a phenomenon found with a wide range of proteins, was overcome by immobilization of polyethylene oxide chains. A general approach was then advanced to enable the selective recognition and binding of target proteins by conjugation of their specific receptors to polyethylene oxide-functionalized nanotubes. This scheme, combined with the sensitivity of nanotube electronic devices, enabled highly specific electronic sensors for detecting clinically important biomolecules such as antibodies associated with human autoimmune diseases. This work led to two important directions of study. The first is the utilization of nanotubes in detecting serum proteins, including disease markers, auto-antibodies, and antibodies (e.g. after therapeutic interventions or vaccinations). The second is the synthesis and fabrication of high-density nanotube device microarrays [104, 194] for proteomics applications, aimed at detecting large numbers of different proteins in a multiplex fashion by using purely electrical transducers. These arrays are attractive because no labeling is required and all aspects of the assay can be carried out in solution phase.

Specific peptides with selective affinity for carbon nanotubes via phage display have been discovered [260]. Consensus binding sequences showed a motif rich in

Histidine and Tryptophane amino acids. Analysis of the hydrophobicity of the peptide chains suggested that they act as symmetric detergents, with a hydrophobic region in the middle and hydrophilic regions at the ends. Binding specificity has been confirmed by demonstrating direct attachment of nanotubes to phage and free peptides immobilized on microspheres. Different possible organizations of surfactant molecules chemically adsorbed on the surface of CNTs can be envisioned [200]. The molecules can be oriented perpendicularly to the surface of the nanotube, forming a monolayer; alternatively, they may be organized into half-cylinders on the surface of the tubes, either oriented parallel or perpendicular to the tube axis. These findings, taken together, open the possibility that CNTs may be functionalized with short peptides for development of biosensors or biofilters, according to the possible peptide orientation on CNTs surface.

In conclusion, CNT surface can be functionalized with proteins, lipids, DNA and other biomolecules, and interfacing novel nanomaterials with biological systems could lead to important applications in disease diagnosis, proteomics and nanobiotechnology in general.

7.5 A Glimpse of the Future

In recent years, it has been recognized that bacteriophages have several potential applications in the modern biotechnology industry. They have been proposed, aside from the above described detection of pathogenic bacteria, as alternatives to antibiotics; as delivery vehicles for protein and DNA vaccines; as gene therapy delivery vehicles and as tools for screening libraries of proteins, peptides or antibodies. This diversity and the ease of their manipulation and production means that they have potential uses in research, therapeutics and manufacturing in both the biotechnology and medical fields.

Following their discovery and initial characterization in the early twentieth century, there was much interest in phages once their potential as antibacterial agents was realized. However, following the discovery and general application of antibiotics in the 1930s and 1940s, interest in the practical and therapeutic uses of bacteriophages waned. In the 1950s, bacteriophage research underwent something of a revival, with phages such as lambda and the T-even series being studied as model systems in the newly emerging field of molecular genetics. Since then, phages have continued to be studied but, with a few exceptions, they have been mainly seen as research tools. Only recently, there has been a renewed interest in the applied use of bacteriophages in a diverse range of fields, including their use in phage display, as anti-bacterial agents (phage therapy), in the development of phage-delivered vaccines, as delivery vehicles for gene therapy and the use of a specific phage for bacterial typing. Some of these applications are reviewed by Clark and March [63].

Phage therapy involves the use of lytic phage to specifically kill pathogenic bacteria as an alternative to antibiotics. It is now generally accepted that a lack of understanding of phage biology and poor quality controls when preparing therapeutic stocks contributed to unreliable and inconsistent results in many of the early phage-therapy trials [148, 236]: in some cases, phage preparations for therapeutic use were found to contain few or no viable particles. Recently, it has become apparent that if some of the problems initially encountered with phage therapy can be overcome, it might have potential uses as an alternative or addition to antibiotic therapy. Phage therapy has been tested in humans, animals and plants, with varying degrees of success, and has also been proposed as a means to decontaminate carcasses and the environment. Phage have potential advantages but also several disadvantages when compared with antibiotics; thus, it seems unlikely that phage therapy will ever replace antibiotics. However, with the increasing incidence of antibiotic-resistant bacteria, there is a clear potential for it to be used in a complementary fashion, particularly in cases where phages can be applied externally and are, therefore, less likely to be removed by the immune system.

Phage-display libraries can be screened in several ways to isolate displayed peptides or proteins with practical applications [20, 259, 267]. For example, it is possible to isolate displayed peptides binding target proteins with affinities similar to those of antibodies, which can then be used as therapeutics that act either as agonists or through the inhibition of receptor–ligand interactions. Furthermore, phage-displayed peptides may be used as *signal peptides* able to trigger complex cell responses. Studies are in progress in our laboratory on phage-display selection of peptides that mimic ligands of cell receptors involved in modulating cell processes such as proliferation, apoptosis and differentiation, for their potential applications, respectively, in regenerative medicine, anti-tumoral development and stem cell differentiation.

Phages have been used as potential vaccine delivery vehicles in two different ways: by directly vaccinating with phages carrying vaccine antigens on their surface or by using the phage particle to deliver a DNA vaccine expression cassette that has been incorporated into the phage genome [61, 62]. In phage-display vaccination, phages can be designed to display a specific antigenic peptide or protein on their surface. Alternatively, phages displaying peptide libraries can be screened with a specific antiserum to isolate novel protective antigens or mimetopes – peptides that mimic the secondary structure and antigenic properties of a protective carbohydrate, protein or lipid, despite having a different primary structure [103, 192]. The serum of convalescents can also be used to screen phage-display libraries to identify potential vaccines against a specific disease, without prior knowledge of protective antigens [172]. Rather than generating a transcriptional fusion to a coat protein, substances can also be artificially conjugated to the surface of phages after growth, which further increases the range of antigens that can be displayed [178]. More recently, it has also been shown that unmodified phages can be used to deliver DNA vaccines more efficiently than standard plasmid DNA vaccination [61, 62, 132, 166]. The vaccine gene, under the control of an eukaryotic expression cassette, is cloned into a standard lambda bacteriophage, and purified whole phage particles are injected into the host. The phage coat protects the DNA from degradation and, because it is a virus-like particle, it should target the vaccine to the antigen presenting cells.

One particularly novel use for phage-displayed peptides is in targeted therapy. One example was in the development of a nasally delivered treatment against cocaine addiction [86]: whole phage particles delivered nasally can enter the central nervous system where a specific phage-displayed antibody can bind to cocaine molecules and prevent their action on the brain. Theoretically, it might also be possible to modify the surface of a bacteriophage by incorporating specific protein sequences to preferentially target the particle to particular cell types, e.g. galactose residues to target galactose-recognizing hepatic receptors in the liver [178] or peptides isolated by screening phage-display libraries to target dendritic [74] or Langerhans cells [173]. To screen phages for the ability to target specific tissue types, phage-display libraries have been passaged through mice several times and at each stage, phages were isolated from specific tissues [198]. A similar in vitro screening strategy was also used to isolate phages displaying peptides that showed increased cytoplasmic uptake into mammalian cells [131]. Phage-displayed peptides so selected may be used as targeted vehicles for antibiotics or anti-tumorals, or act themselves as targeted anti-bacterials and anti-tumorals. Specific phage-displayed peptides could be used, for example, in anticancer therapy either directly inducing apoptosis processes or targeting anti-tumorals to cancer cells, or also targeting microorganisms that, in turn, specifically infect tumorals cells.

Finally, phages have also been proposed as potential therapeutic gene delivery vectors [13, 92]. The phage coat protects the DNA from degradation after injection, and the ability to display foreign molecules on the phage coat also enables targeting of specific cell types, a prerequisite for effective gene therapy. Both artificial covalent conjugation [151] and phage display [152] have been used to display targeting and/or processing molecules on the phage surface. This demonstrates, again, the versatility of phages, showing that tissue targeting can be achieved either by rational design or by the screening of random phage-display libraries.

References

1. Adey NB, Mataragnon AH, Rider JE, Carter JM, Kay BK (1995) Characterization of phage that bind plastic from phage-displayed random peptide libraries. Gene 156:27–31
2. Allison DG, Sutherland IW (1987) Role of exopolysaccharides in adhesion of freshwater bacteria. J Gen Microbiol 133:1319–1327
3. An YH, Friedman RJ (1998) Concise review of mechanisms of bacterial adhesion to biomaterial surfaces J Biomed Mat Re 43:338–348
4. Andersson H, van den Berg A (2003) Microfluidic devices for cellomics: A review. Sensors and Actuators B Chem 92:315–325
5. Arap W, Kolonin MG, Trepel M, Lahdenranta J, Cardo-Vila M, Giordano RJ, Mintz PJ, Ardelt PU, Yao VJ, Vidal CI, Chen L, Flamm A, Valtanen H, Weavind LM, Hicks ME, Pollock RE, Botz GH, Bucana CD, Koivunen E, Cahill D, Troncoso P, Baggerly KA, Pentz RD, Do KA, Logothetis CJ, Pasqualini R (2002) Steps toward mapping the human vasculature by phage display. Nature Med 8:121–127
6. Badjic JD, Balzani V, Credi A, Silvi S, Stoddart JF (2004) A molecular elevator. Science 303:1845–1849

7. Bak P, Sneppen K (1993) Punctuated equilibrium and criticality in a simple model of evolution. Phys Rev Lett 71:4083–4086
8. Bak P, Tang C (1989) Earthquakes as a self-organized critical phenomena. J Geophys Res 94:15635–15637
9. Bak P, Tang C, Wiesenfeld K (1988) Self-organized criticality. Phys Rev A 38:364–374
10. Balass M, Heldman Y, Cabilly S, Givol D, Katchalski-Katzir E, Fuchs S (1993) Identification of a hexapeptide that mimics a conformation-dependent binding site of acetylcholine receptor by use of a phage-epitope library. Proc Natl Acad Sci USA 90:10638–10642
11. Balasubramanian K, Burghard M (2006) Biosensors based on carbon nanotubes [Review]. Anal Bioanal Chem 385:452–468
12. Balavoine F, Schultz P, Richard C, Mallouh V, Ebbesen TW, Mioskowski C (1999) Helical crystallization of proteins on carbon nanotubes. Angew Chem Int Ed 38:1912–1915
13. Barry MA, Dower WJ, Johnston SA (1996) Toward cell-targeting gene therapy vectors: Selection of cell-binding peptides from random peptide-presenting phage libraries. Nat Med 2:299–305
14. Bashir R (2004) BioMEMS: state-of-the-art in detection, opportunities and prospects. Adv Drug Del Rev 56:1565–1586
15. Battifora H (1986) The multitumor (sausage) tissue block: novel method for immunohistochemical antibody testing. Lab Invest 55:244–248
16. Baughman RH, Zakhidov AA, de Heer WA (2002) Carbon Nanotubes – The route towards applications. Science 297:787–792
17. Belkin S, Smulski DR, Dadon S, Vollmer AC, Van Dyk TK, Larossa RA (1997) A panel of stress-responsive luminous bacteria for the detection of selected classes of toxicants. Water Research 31:3009–3016
18. Ben Jacob E, Schochet O, Tenenbaum A, Cohen I, Czirok A, Tamas V (1994) Generic modeling of cooperative growth patterns in bacterial colonies. Nature 368:46–49
19. Ben Jacob E, Cohen I, Shochet O, Tenenbaum A, Czirok A, Vicsek T (1995) Cooperative formation of chiral patterns during growth of bacterial colonies. Phys Rev Lett 75:2899–2902
20. Benhar I (2001) Biotechnological applications of phage and cell display. Biotechnol Adv 19:1–33
21. Berglund J, Lindbladh C, Nicholls IA, Mosbach K (1998) Selection of phage display combinatorial library peptides with affinity for a yohimbine imprinted methacrylate polymer. Anal Commun 35:3–7
22. Berlekamp ER, Conway JH, Guy RK (1982) In: Winning ways for your mathematical plays. Academic Press, New York
23. Bershadsky 1996 Bershadsky, A., Chausovsky, A., Becker, E., Lyubimova, A. and Geiger, B. (1996). Involvement of microtubules in the control of adhesion-dependent signal transduction. Curr Biol 6:1279–1289
24. Bishop-Hurley SL, Schmidt FJ, Erwin AL, Smith AL (2005) Peptides selected for binding to a virulent strain of *Haemophilus influenzae* by phage display are bactericidal. Antimicrob Ag Chemother 49:2972–2978
25. Bond DR, Lovley DR (2003) Electricity production by Geobacter sulfurreducens attached to electrodes. Appl Environ Microbiol 69:1548–1555
26. Boretos JW, Eden M (1984) Contemporary Biomaterials, Material and Host Response, Clinical Applications, New Technology and Legal Aspects. Noyes Publications, Park Ridge, NJ, pp 232–233
27. Bos R, van der Mei HC, Busscher HJ (1995) A quantitative method to study co-adhesion of microorganisms in parallel plate flow chamber. II: Analysis of the kinetics of co-adhesion. J Microbiol Methods 23:169–182
28. Bos R, van der Mei HC, Busscher HJ (1999) Physico-chemistry of initial microbial adhesive interactions – its mechanisms and methods for study. FEMS Microbiol Rev 23:179–230
29. Bousse L (1996) Whole cell biosensors. Sens Actuators B Chem B 34:270–275
30. Broach JR, Thorner J (1996) High-throughput screening for drug discovery. Nature 384: 14–16

31. Brown S (1992) Engineered iron oxide-adhesion mutants of the *Escherichia coli* phage l receptor. Proc Natl Acad Sci USA 89:8651–8655
32. Brown S (1997) Metal-recognition by repeating polypeptides. Nature Biotechnol 15: 269–272
33. Bruchez M, Moronne M, Gin P, Weiss S, Alivisatos AP (1998) nanocrystals as fluorescent biological labels. Science 281:2013–2016
34. Buchanan M (2000) Ubiquity. Weidenfeld & Nicholson eds, London
35. Bucolo M, Carnazza S, Fortuna L, Frasca M, Guglielmino S, Marletta G (2004) Self-organization and emergent models in bacterial adhesion on engineered polymer surfaces. International Symposium on Circuits and Systems (ISCAS2004) III:689–692
36. Bullen RA, Arnot TC, Lakeman JB, Walsh FC (2005) Biofuel cells and their development. Bios Bioelectr 21:2015–2045
37. Buret A, Ward KH, Olson ME, Costerton JW (1991) An in vivo model to study the pathobiology of infectious biofilms on biomaterial surfaces. J Biomed Mater Res 25:865–874
38. Burns MA, Johnson BN, Brahmasandra SN, Handique K, Webster JR, Krishnan M, Sammarco TS, Man P, Jones D, Heldsinger D, Mastrangelo CH, Burke DT (1998) Integrated nanoliter DNA analysis device. Science 282:484–487
39. Burridge K, Turner CE, Romer LH (1992) Tyrosine phosphorylation of paxillin and pp125FAK accompanies cell adhesion to extracellular matrix : A role in cytoskeletal assembly. J Cell Biol 119:893–903
40. Busscher HJ, Bos R, van der Mei HC (1995) Initial microbial adhesion is a determinant for the strength of biofilm adhesion. FEMS Microbiol Lett 128:229–234
41. Carlson JM, Langer JS (1989) Properties of earthquakes generated by fault dynamics. Phys Rev Lett 62:2632–2635
42. Carnazza S, Satriano C, Guglielmino S, Marletta G (2005a) Fast exopolysaccharide secretion of *Pseudomonas aeruginosa* on polar polymer surfaces. J Coll Interf Sci 289:386–393
43. Carnazza S, Satriano C, Marletta G, Frasca M, Fortuna L, Guglielmino S (2005b) Auto-organizzazione di cellule microbiche sottoposte a dewetting su superfici polimeriche modificate: nuove prospettive nel patterning cellulare. II Simposio sulle Tecnologie Avanzate. Applicazioni delle Nanotecnologie per la Difesa nei Settori Strutturale, Elettronico, Biotecnologico. Ministero della Difesa, Segredifesa, Roma, Italy
44. Carnazza S, Satriano C, Marletta G, Guglielmino S (2005c) A novel surface polymer modification to improve controlled cell adhesion and to prevent bacterial colonisation. 7th FISV Symposium, Riva del Garda, Italy
45. Carnazza S, Satriano C, Guglielmino S (2006a) Self-organization of yeast cells on modified polymer surfaces after dewetting: New perspectives in cellular patterning. J Physics Condensed Matter 18: S33 Nanoscience and Nanotechnology:S2221–S2230
46. Carnazza S., Satriano C., Marletta G., Frasca M., Fortuna L., Guglielmino S. (2006b) Nuove prospettive di preparazione di array microbici come piattaforma per "lab-on-a-cell" chip e microsistemi diagnostici" III Simposio sulle Tecnologie Avanzate. Utilizzo e Applicazioni. Ministero della Difesa, Segredifesa, Roma, Italy
47. Carnazza S, Caratozzolo M, Fortuna L, Frasca M, Guglielmino S, Guerrieri G, Marletta G (2007a) Self-organizing models of bacterial aggregation states. XIV International Conference on Waves and stability in continuous media, Scicli (Ragusa), Italy
48. Carnazza S, Gioffrè G, Felici F, Guglielmino S (2007b) Recombinant phage probes for *Listeria monocytogenes*. J Phys Condensed Matter 19:395011
49. Cerrai P, Guerra GD, Tricoli M, Krajewski A, Ravaglioli A, Martinetti R, Dolcini L, Fini M, Scarano A, Piattelli A (1999) Periodontal membranes from composites of hydroxyapatite and bioresorbable block copolymers. J Mater Sci Mater Med 10:677–682
50. Characklis WG, Marshall KC (Eds) (1990) Biofilms. John Whiley & Sons, New York
51. Chen K, Bak P (1989) Is the universe operating at a self-organized critical state? Phys Lett A 140:299–302
52. Chen RJ, Zhang Y, Wang D, Dai H (2001) Noncovalent sidewall functionalization of single-walled carbon nanotubes for protein immobilization. J Am Chem Soc 123:3838–3839

53. Chen RJ, Bangsaruntip S, Drouvalakis KA, Wong Shi Kam N, Shim M, Li Y, Kim W, Utz PJ, Dai H (2003) Noncovalent functionalization of carbon nanotubes for highly specific electronic biosensors. Proc Natl Acad Sci USA 100:4984–4989
54. Chen T, Barton SC, Binyamin G, Gao Z, Zhang Y, Kim H-H, Heller A (2001). A miniature Biofuel Cell. J Am Chem Soc 123:8630–8631
55. Chen W-T, Singer SJ (1982) Immunoelectron microscopic studies of the sites of cell-substratum and cell-cell contacts in cultured fibroblasts. J Cell Biol 95:205–222
56. Chen W-T, Hasegawa E, Hasegawa T, Weinstock C, Yamada KM (1985) Development of cell surface linkage complexes in cultured fibroblasts. J Cell Biol 100:1103–1114
57. Chiao M, Lam KB, Lin L (2003) Micromachined microbial fuel cells. In: Proc IEEE Micro Electro Mechanical Systems (MEMS), pp 383–386
58. Chiao M, Lam KB, Lin L (2006) Micromachined microbial and photosynthetic fuel cells. J Micromech Microeng 16:2547–2553
59. Chu PK, Chen JY, Wang LP, Huang N (2002) Plasma-surface modification of biomaterials. Mat Sci Eng R 36:143–206
60. Clark EA, Brugge JS (1995) Integrins and signal transduction pathways: the road taken. Science 268:233–239
61. Clark JR, March JB (2004a) Bacterial viruses as human vaccines? Expert Rev Vaccines 3:463–476
62. Clark JR, March JB (2004b) Bacteriophage-mediated nucleic acid immunization. FEMS Immunol Med Microbiol 40:21–26
63. Clark JR, March JB (2006) Bacteriophages and biotechnology: vaccines, gene therapy and antibacterials. Trends Biotech 24:212–218
64. Clark P, Connolly P, Curtis ASG, Dow JAT., Wilkinson, CDW (1990) Topographical control of cell behaviour II. Multiple grooved substrata. Development 108:635–644
65. Clark P, Connolly P, Curtis ASG, Dow JAT, Wilkinson CDW (1991) Cell guidance by ultrafine topography in vitro. J Cell Sci 99:73–77
66. Clark P, Connolly P, Moores GR (1992) Cell guidance by micropatterned adhesiveness in vitro. J Cell Sci 103:287–292
67. Cortese R, Monaci P, Nicosia A, Luzzago A, Felici F, Galfre G, Pessi A, Tramontano A, Sollazzo M (1995) Identification of biologically active peptides using random libraries displayed on phage. Curr Opin Biotechnol 6:73–80
68. Cowan SE, Lipmann D, Keasling JD (2001) Development of engineered biofilms on poly-L-lysine patterned surfaces. Biotechnology Letters 23:1235–1241
69. Craighead HG, James CD, Turner AMP (2001) Chemical and topographical patterning for directed cell attachment. Curr Opin Sol State Mat Sci 5:177–184
70. Cui X, Hetke JF, Wiler JA, Anderson DJ, Martin DC (2001) Electrochemical deposition and characterization of conducting polymer polypyrrole/PSS on multichannel neural probes. Sensors Actuat A 93:8–18
71. Cui Y, Wei Q, Park H, Lieber CM (2001) Nanowire nanosensors for highly sensitive and selective detection of biological and chemical species. Science 293 :1289–1292
72. Cunliffe D, Smart CA, Alexander C, Vulfson EN (1999) Bacterial adhesion at synthetic surfaces. Appl Environ Microbiol 65:4995–5002
73. Cuong AP, Jung SJ, Phung NT, Lee J, Chang IS, Kim BH, Yi H, Chun J (2003) A novel electrochemically active and Fe(III)-reducing bacterium phylogenetically related to *Aeromonas hydrophila*, isolated from a microbial fuel cell. FEMS Microbiol Lett 223:129–134
74. Curiel TJ, Morris C, Brumlik M, Landry SJ, Finstad K, Nelson A, Joshi V, Hawkins C, Alarez X, Lackner A, Mohamadzadeh M (2004) Peptides identified through phage display direct immunogenic antigen to dendritic cells. J Immunol 172:7425–7431
75. Curtis A, Riehle M (2001) Tissue engineering: The biophysical background. Phys Med Biol 46:R47–65
76. Curtis ASG, Wilkinson CDW (2001) Nanotechniques and approaches in biotechnology. Trends Biotechnol 19:97–101

77. Curtis ASG, Casey B, Gallagher JO, Pasqui D, Wood MA, Wilkinson CDW (2001) Substratum nanotopography and the adhesion of biological cells. Are symmetry or regularity of nanotopography important? Bioph Chem 94:275–283
78. Cwirla SE, Peters EA, Barrett RW, Dower WJ (1990) Peptides on phage: A vast library of peptides for identifying ligands. Proc Natl Acad Sci USA 87:6378–6382
79. Dalton BA, Evans MDM, McFarlands GA, Steele JG (1999) Modulation of corneal epithelial stratification by polymer surface topography. J Biomed Mater Res 45:384–394
80. Dankert J, Hogt AH, Feijen J (1986) Biomedical polymers: bacterial adhesion, colonization, and infection. CRC Crit Rev Biocompat 2:219–301
81. Davis JJ, Green M, Hill A, Leung Y, Sadler PJ, Sloan J, Xavier A, Tsang SC (1998) The immobilization of proteins in carbon nanotubes. Inorg Chim Acta 272:261–266
82. Dee KC, Puleo DA, Bizios R (2002) An introduction to tissue-biomaterial interactions. John Wiley & Sons, New York
83. Deheyder B, Overmeire A, Van Langenhove H, Verstraete W (1994) Ethene removal from a synthetic waste gas using a dry biobed. Biotechnol Bioeng 44:642–648
84. Delaney GM, Bennetto HP, Mason JR, Roller HD, Stirling JL, Thurston CF (1984) Electron-transfer coupling in microbial fuel cells: 2. Performance of fuel cells containing selected micoorganism-mediator-substrate combinations. J Chem Tech Biotechnol 34B:13–27
85. Denkov ND, Velev OD, Kralchevsky PA, Ivanov IB, Yoshimura H, Nagayama K (1992) Mechanism of formation of two-dimensional crystals from latex particles on substrates. Langmuir 8:3183–3190
86. Dickerson TJ, Kaufmann GF, Janda KD (2005) Bacteriophage-mediated protein delivery into the central nervous system and its application in immunopharmacotherapy. Expert Opin Biol Ther 5:773–781
87. Donlan RM (2001) Biofilm formation: a clinically relevant microbiological process. Clin Infect Dis 338:1387–1392
88. Dragt AJ, van Ham J (eds) (1992) Biotechniques for air pollution abatement and odour control policies. Elsevier Biomedical Press, Amsterdam
89. Dresselhaus MS, Dresselhaus G, Eklund PC (1996) In: Science of Fullerenes and Carbon Nanotubes. Academic Press, San Diego
90. Dresselhaus MS, Dresselhaus G, Avouris P (2001) In: Carbon Nanotubes. Springer, Berlin
91. Du C, Cui FZ, Zhu XD, de Groot K (1999) Three-dimentional nano- HAp/collagen matrix loading with osteogenic cells in organ culture. J Biomed Mater Res 44:407–415
92. Dunn IS (1996) Mammalian cell binding and transfection mediated by surface-modified bacteriophage lambda. Biochimie 78:856–861
93. Eginton PJ, Gibson H, Holah J, Handley PS, Gilbert P (1995) The influence of substratum properties on the attachment of bacterial cells. Colloids Surf B Biointerfaces 5:153–159
94. Elsgaard L (1998) Ethylene removal by a biofilter with immobilized bacteria. Appl Env Microbiol 64:4168–4173
95. Emili AQ, Cagney G (2000) Large-scale functional analysis using peptide or protein arrays. Nature Biotechnol 18:393–397
96. Entzeroth M (2003) Emerging trends in high-throughput screening. Curr Opin Pharm 3: 522–529
97. Erlanger BF, Chen B, Zhu M, Brus LE (2001) Binding of an anti-fullerene IgG monoclonal antibody to single wall carbon nanotubes. Nano Lett 1:465–467
98. Felici F, Castagnoli L, Musacchio A, Jappelli R, Cesareni G (1991) Selection of antibodies ligands from a large library of oligopeptides expressed on a multivalent exposition vector. J Mol Biol 222:301–310
99. Felici F, Luzzago A, Monaci P, Nicosia A, Sollazzo M, Traboni C (1995) Peptide and protein display on the surface of filamentous bacteriophage. In: El-Gewely MR (ed) Biotechnology Annual Review, Elsevier Science BV, Amsterdam, vol 1, pp 149–183
100. Fernandes PB (1998) Technological advances in high-throughput screening. Curr Opin Chem Biol 2:597–603

101. Flynn CE, Mao C, Hayhurst A, Williams JL, Georgiou G, Iverson B, Belcher AM (2003) Synthesis and organization of nanoscale II VI semiconductor materials using evolved peptide specificity and viral capsid assembly. J Mater Chem 13:2414–2421
102. Fodor SPA, Read JL, Pirrung MC, Stryer L, Lu AT, Solas D (1991) Light-directed, spatially addressable parallel chemical synthesis. Science 251:767–773
103. Folgori A, Tafi R, Meola A, Felici F, Galfre G, Cortese R, Monaci P, Nicosia A (1994) A general strategy to identify mimotopes of pathological antigens using only random peptide libraries and human sera. EMBO (Eur Mol Biol Organ) J 13:2236–2243
104. Franklin NR, Li Y, Chen RJ, Javey A, Dai H (2001) Patterned growth of singlewalled carbon nanotubes on full 4-inch wafers. Appl Phys Lett 79:4571–4573
105. Fritz J, Baller MK, Lang HP, Rothuizen H, Vettiger P, Meyer E, Guntherodt H-J, Gerber Ch, Gimzewski JK (2000) Translating biomolecular recognition into nanomechanics. Science 288:316–318
106. Fung ET, Thulasiraman V, Weinberger SR, Dalmasso EA (2001) Protein biochips for differential profiling. Curr Opin Biotechnol 12:65–69
107. Gabriel D, Deshusses MA (2003) Retrofitting existing chemical scrubbers to biotrickling filters for H_2S emission control. Proc Natl Acad Sci USA 100:6308–6312
108. Geiger B (1979) A 130K protein from chicken gizzard: Its localization at the termini of microfilament bundles in cultured chicken cells. Cell 18:193–205
109. Giamblanco N, Satriano C, Carnazza S, Guglielmino S, Marletta G (2005) Fibroblast adhesion onto single and binary proteins films adsorbed on polycrystalline gold. 19th European Symposium on Biomaterials (ESB 2005), Sorrento, Italy
110. Gierak J, Mailly D, Faini G, Pelouard JL, Denk P, Pardo F, Marzin JY, Septier A, Schmid G, Ferre J, Hydman R, Chappert C, Flicstein J, Gayral B, Gerard JM (2001) Nano-fabrication with focused ion beams. Microelectr Engin 57:865–875
111. Gil G, Chang I, Kim B, Kim M, Jang J, Park H, Kim H (2003) Operating parameters affecting the performance of a mediator-less microbial fuel cell. Biosens Bioelectron 18:327–334
112. Gingell D (1993) In: Jones G, Wigley C, Wran R (eds) SEB Symposium, vol 47, pp 1–33
113. Giordano RJ, Cardo-Vila M, Lahdenranta J, Pasqualini R, Arap W (2001) Biopanning and rapid analysis of selective interactive ligands. Nature Med 7:1249–1253
114. Gorby YA et al. (2006) Electrically conductive bacterial nanowires produced by *Shewanella oneidensis* strain MR-1 and other microorganisms. Proc Natl Acad Sci USA 103:11358–11363
115. Grant RP, Spitzfaden C, Altroff H, Campbell ID, Mardon HJ (1997) Structural requirements for biological activity of the ninth and tenth FIII domains of human fibronectin. J Biol Chem 272:6159–6166
116. Gray C, Boyde A, Jones SJ (1996) Topographically induced bone formation in vitro: implications for bone implants and bone grafts. Bone 18:115–123
117. Gregory KB, Bond DR, Lovley DR (2004) Graphite electrodes as electron donors for anaerobic respiration. Environ Microbiol 6:596–604
118. Gregory KB, Lovley DR (2005) Remediation and recovery of uranium from contaminated subsurface environments with electrodes. Environ Sci Technol 39:8943–8947
119. Hall BL, Smit-McBride Z, Privalsky ML (1993) Reconstitution of the retinoid X receptor function and combinatorial regulation of other nuclear hormone receptors in the yeast *Saccharomyces cerevisiae*. Proc Natl Acad Sci USA 90:6929–6933
120. Hansen KM, Ji HF, Wu G, Datar R, Cote R, Majumdar A, Thundat T (2001) Cantilever-based optical deflection assay for discrimination of DNA single nucleotide mismatches. Anal Chem 73:1567–1571
121. Hartmans S, de Bont JAM, Harder W (1989) Microbial metabolism of short-chain unsaturated hydrocarbons. FEMS Microbiol Rev 63:235–264
122. Heller MJ (1996) An active microelectronics device for multiplex DNA analysis. IEEE Eng Med Biol Mag 15:100–104
123. Hermansson M (1999) The DLVO theory in microbial adhesion. Colloids Sur B Biointerfaces 14:105–119

124. Hertzberg RP, Pope AJ (2000) High-throughput screening: New technology for the 21st century. Curr Opin Chem Biol 4:445–451
125. Horiuchi KY, Wang Y, Diamond SL, Ma H (2006) Microarrays for the functional analysis of the chemical-kinase interactome. J Biomol Screen 11:48–56
126. Huang TS, Tzeng Y, Liu YK, Chen YC, Walker KR, Guntupalli R, Liu C (2004) Immobilization of antibodies and bacterial binding on nanodiamond and carbon nanotubes for biosensor applications. Diam Rel Mat 13:1098–1102
127. Huang W, Taylor S, Fu K, Lin Y, Zhang D, Hanks TW, Rao A, Sun Y (2002) Attaching proteins to carbon nanotubes via diimide-activated amidation. Nano Lett 2:311–314
128. Hutmacher DW (2001) Scaffold design and fabrication technologies for engineering tissues-state of the art and future perspectives. J Biomat Sci-Polym E 12:107–124
129. Hynes RO (1987) Integrins: a family of cell surface receptors. Cell 48:549–554
130. Hynes RO (1992) Integrins: Versatility, modulation, and signaling in cell adhesion. Cell 69:11–25
131. Ivanenkov VV, Menon AG (2000) Peptide-mediated transcytosis of phage display vectors in MDCK cells. Biochem Biophys Res Commun 276:251–257
132. Jepson CD, March JB (2004) Bacteriophage lambda is a highly stable DNA vaccine delivery vehicle. Vaccine 22:2413–2419
133. Jockusch BM, Bubeck P, Giehl K, Kroemker M, Moschner J, Rothkegel M, Rudiger M, Schluter K, Stanke G, Winkler J (1995) The molecular architecture of focal adhesions. Annu Rev Cell Dev Biol 11:379–416
134. Joshi KA, Tang J, Haddon R, Wang J, Chen W, Mulchandani A (2005) A disposable biosensor for organophosphorus nerve agents based on carbon nanotubes modified thick film strip electrode. Electroanalysis 17:54–58
135. Julian W, Wimpenny T, Colasanti R (1997) A unifying hypothesis for the structure of microbial biofilms based on cellular automaton models. FEMS Microbiol Ecol 22:1–16
136. Juliano RL, Haskill S (1993) Signal transduction from the extracellular matrix. J Cell Biol 120:577–585
137. Kam Z, Volberg T, Geiger B (1995) Mapping of adherence juction components using microscopic resonance energy-transfer imaging. J Cell Sci 108:1051–1062
138. Kane RS, Takayama S, Ostuni E, Ingber DE, Whitesides GM (1999) Patterning proteins and cells using soft lithography, Biomaterials 20:2363–2376
139. Katz B-Z, Zamir E, Bershadsky A, Kam Z, Yamada KM, Geiger B (2000) Physical state of the extracellular matrix regulates the structure and molecular composition of cell-matrix adhesions. Mol Biol Cell 11:1047–1060
140. Kauffman S, Johnson S (1992) In: Langton CG, Taylor C, Farmer JD, Rasmussen S (eds) Artificial Life II. Addison-Wesley, Reading (Mass.), Redwood City, pp 325–369
141. Kessler D, Levine H (1993) Pattern formation in Dictyostelium via the dynamics of cooperative biological entities. Phys Rev E 48:4801–4804
142. Kim BH, Kim HJ, Hyun MS, Park DH (1999) Direct electrode reaction of Fe(III)-reducing bacterium *Shewanella putrefacience*. J Microbiol Biotechnol 9:127–131
143. King K, Dohlman HG, Thorner J, Caron MG, Lefkowitz RJ (1990) Control of yeast mating signal transduction by a mammalian b2-adrenergic receptor and Gsa subunit. Science 250:121–123
144. Klein RD, Geary TG (1997) Recombinant microorganisms as tools for high throughput screening for non-antibiotic compounds. J Biomol Screen 2:41–49
145. Kononen J, Bubendorf L, Kallioniemi A, Barlund M, Schraml P, Leighton S, Torhorst J, Mihatsch MJ, Sauter G, Kallioniemi OP (1998) Tissue microarrays for high-throughput molecular profiling of tumor specimens. Nature Med 4:844–847
146. Konturri K, Pentti P, Sundholm G (1998) Polypyrrole as a model membrane for drug delivery. J Electroanal Chem 453:231–238
147. Kralli A, Bohen SP, Yamamoto KR (1995) LEM1, an ATP-binding cassette transporter, selectively modulates the biological potency of steroid hormones. Proc Natl Acad Sci USA 92:4701–4705

148. Kutter E, Sulakvelidze A (2005) Bacteriophages biology and applications. CRC Press, Boca Raton, Florida
149. Langer R, Tirrell DA (2004) Designing materials for biology and medicine. Nature 428:487–492
150. Langer R, Vacanti JP (1993) Tissue engineering. Science 260:920–926
151. Larocca D, Witte A, Johnson W, Pierce GF, Baird A (1998) Targeting bacteriophage to mammalian cell surface receptors for gene delivery. Hum Gene Ther 9:2393–2399
152. Larocca D, Kassner PD, Witte A, Ladnera RC, Pierce GF, Baird A (1999) Gene transfer to mammalian cells using genetically targeted filamentous bacteriophage. FASEB J 13:727–734
153. LaVan DA, Lynn DM, Langer R (2002) Moving smaller in drug discovery and delivery. Nat Rev Drug Discov 1:77–84
154. Lee LT, da Silva MDCV, Galembeck F (2003) Dewetting patterns of thin films of charged polymer solutions. Langmuir 19:6717–6722
155. Lee LT, Leite CAP, Galembeck F (2004) Controlled nanoparticle assembly by dewetting of charged polymer solutions. Langmuir 20:4430–4435
156. Lenihan JS, Gavalas VG, Wang J, Andrews R, Bachas LG (2004) Protein immobilization on carbon nanotubes through a molecular adapter. J Nanosci Nanotechnol 4:600–604
157. Liao J-D, Wang S-H, Hsu D-J (2001) Studies on the early detection of wastewater's toxicity using a microbial sensing system. Sens Actuators B 72:167–173
158. Lithgow AM, Romero L, Sanchez IC, Souto FA, Vega CA (1986) Interception of electron-transport chain in bacteria with hydrophilic redox mediators. J Chem Res (S):178–179
159. Liu H, Grot S, Logan BE (2005) Electrochemically assisted microbial production of hydrogen from acetate. Environ Sci Technol 39:4317–4320
160. Luppa PB, Sokoll LJ, Chan DW (2001) Immunosensors-principles and applications to clinical chemistry (Review). Clin Chim Acta 314:1–26
161. Luzzago A, Felici F (1998) Construction of disulfide-constrained random peptide libraries displayed on phage coat protein VIII. Methods Mol Biol 87:155–164
162. Ma H, Horiuchi KY (2006) Chemical Microarray: a new tool for drug screening and discovery. Drug Discovery Today 11:661–668
163. Madou M (1997) Fundamentals of microfabrication. CRC Press, New York
164. Madou MJ, Lee LJ, Daunert S, Lai I, Chih-Hsin S (2001) Design and fabrication of CD-like microfluidic platforms for diagnostics: microfluidic functions. Biomed Microdevices 3:245–254
165. Magnuson TS, Isoyama N, Hodges-Myerson AL, Davidson G, Maroney MJ, Geesey GG, Lovley DR (2001) Isolation, characterization and gene sequence analysis of a membrane-associated 89 kDa Fe(III) reducing cytochrome c from *Geobacter sulfurreducens*. Biochem J 359:147–152
166. March JB, Clark JR, Jepson CD (2004) Genetic immunization against hepatitis B using whole bacteriophage lambda particles. Vaccine 22:1666–1671
167. Marletta G (1990) Chemical reactions and physical property modifications induced by keV ion beams in polymers. Nucl Instr Meth B 46:295–305
168. Marletta G, Satriano C (2001) Irradiation-controlled adsorption and organization of biomolecules on surfaces: from the nanometric to the mesoscopic level. In: Buzaneva E, Scharff P (eds) Frontiers in molecular-scale science and technology of nanocarbon, nanosilicon and biopolymer multifunctional nanosystems. Kluwer Academic Publishers, Amsterdam, pp 1–13
169. Marletta G, Catalano SM, Pignataro S (1990) Chemical reactions induced in polymers by keV ion, electrons and photons. Surf Interface Sci 16:407–411
170. Marshall KC, Stout R, Mitchell R (1971) Mechanism of the initial events in the sorption of marine bacteria to surfaces. J Gen Microbiol 68:337–348
171. Matsumoto N, Nakasono S, Ohmura N, Saiki H (1999) Extension of logarithmic growth of *Thiobacillus ferrooxidans* by potential controlled electrochemical reduction of Fe(III). Biotechnol Bioeng 64:716–721
172. Mattson MP, Haddon RC, Rao AM (2000) Molecular functionalization of carbon nanotubes and use as substrates for neuronal growth. J Mol Neurosci 14:175–182

173. McGuire MJ, Sykes KF, Samli KN, Timares L, Barry MA, Stemke-Hale K, Tagliaferri F, Logan M, Jansa K, Takashima A, Brown KC, Johnston SA (2004) A library-selected, Langerhans cell targeting peptide enhances an immune response. DNA Cell Biol 23:742–752
174. McKendry R, Zhang J, Arntz Y, Strunz T, Hegner M, Lang HP, Baller MK, Certa U, Meyer E, Güntherodt H-J, Gerber C (2002) Multiple label-free biodetection and quantitative DNA-binding assays on a nanomechanical cantilever array. Proc Natl Acad Sci USA 99: 9783–9788
175. Meola A, Delmastro P, Monaci P, Luzzago A, Nicosia A, Felici F, Cortese R, Galfre G (1995) Derivation of vaccines from mimotopes. Immunologic properties of human hepatitis B virus surface antigen mimotopes displayed on filamentous phage. J Immunol 154:3162–3172
176. Michael KE, Vernekar VN, Keselowsky BG, Meredith JC, Latour RA, Garcìa AJ (2003) Adsorption-Induced conformational changes in fibronectin due to interactions with well-defined surface chemistries. Langmuir 19:8033–8040
177. Min B, Cheng S, Logan BE (2005) Electricity generation using membrane and salt bridge microbial fuel cells. Water Res 39:1675–1686
178. Molenaar TJM, Michon IN, de Haas SA, van Berkel ThJC, Biessen EAL (2002) Uptake and processing of modified bacteriophage M13 in mice: Implications for phage display. Virology 293:182–191
179. Morra M, Cassinelli C (1997) Bacterial adhesion to polymer surfaces: a critical review of surface thermodynamic approaches. J Biomater Sci Polymer Ed 9:55–74
180. Myers JM, Myer CR (2001) Role for outer membrane cytochromes OmcA and OmcB of *Shewanella putrefaciens* MR-1 in reduction of manganese dioxide. Appl Environ Microbiol 67:260–269
181. Neu TR, Marshall KC (1990) Bacterial polymers: Physico-chemical aspects of their interactions at interfaces. J Biomater Appl 5:107–133
182. Neumann AW, Good RJ (1979) Techniques of measuring contact angles. In: Good RJ, Stromberg RR (eds) Surface and colloid science, Plenum, New York, vol 11, pp 31–91
183. Nicolau DV, Taguchi T, Taniguchi H, Yoshikawa S (1999) Negative and positive tone protein patterning on e-beam deep-UV resists. Langmuir 15:3845–3851
184. O'Connor TP, Duerr JS, Bentley D (1990) Pioneer growth cone steering decisions mediated by single filopodial contacts in situ. J Neurosci 10:3935–3946
185. O'Toole GA, Kolter R (1998) Flagellar and twitching motility are necessary for *Pseudomonas aeruginosa* biofilm development. Mol Microbiol 302:295–304
186. Olsen EV, Sorokulova IB, Petrenko VA, Chen I-H, Barbaree JM, Vodyanoy VJ (2006) Affinity-selected filamentous bacteriophage as a probe for acoustic wave biodetectors of *Salmonella typhimurium*. Bios Bioelectr 21:1434–1442
187. Pakalns T, Haverstick K, Fields GB, McCarthy JB, Mooradian DL, Tirrell M (1999) Cellular recognition of synthetic peptide amphiphiles in self-assembled monolayer films. Biomaterials 20:2265–2279
188. Pancrazio JJ, Whelan JP, Borkholder DA, Ma W, Stenger DA (1999) Development and application of cell-based biosensors. Ann Biomed Eng 27:697–711
189. Petrenko VA, Sorokulova IB (2004) Detection of biological threats. A challenge for directed molecular evolution. J Microbiol Methods 58:147–168
190. Petrenko VA, Vodyanoy VJ (2003) Phage display for detection of biological threats. J Microbiol Methods 53:253–562
191. Petrenko VA, Smith GP, Gong X, Quinn T (1996) A library of organic landscapes on filamentous phage. Protein Eng 9:797–801
192. Phalipon A, Folgori A, Arondel J, Sgaramella G, Fortugno P, Cortese R, Sansonetti PJ, Felici F (1997) Induction of anti-carbohydrate antibodies by phage library-selected peptide mimics. Eur J Immunol 27:2620–2625
193. Price LA, Kajkowski EM, Hadcock JR, Ozenberger BA, Pausch MH (1995) Functional coupling of a mammalian somatostatin receptor to the yeast pheromone response pathway. Mol Cell Biol 15:6188–6195

194. Qi P, Vermesh O, Grecu M, Javey A, Wang Q, Dai H, Peng S, Cho KJ (2003) Toward large arrays of multiplex functionalized carbon nanotube sensors for highly sensitive and selective molecular detection. Nano Lett 3:347–351
195. Rabaey K, Boon N, Siciliano SD, Verhaege M, Verstraete W (2004) Biofuel cells select for microbial consortia that self-mediate electron transfer. Appl Environ Microbiol 70:5373–5382
196. Rabaey K, Boon N, Höfte M, Verstraete W (2005) Microbial phenazine production enhances electron transfer in biofuel cells. Environ Sci Technol 39:3401–3408
197. Rajnicek A, McCaig C (1997) Guidance of CNS growth cones by substratum grooves and ridges: effects of inhibitors of the cytoskeleton, calcium channels and signal transduction pathways. J Cell Sci 110:2915–2924
198. Rajotte D, Arap W, Hagedorn M, Koivunen E, Pasqualini R, Ruoslahti E (1998) Molecular heterogeneity of the vascular endothelium revealed by in vivo phage display. J Clin Invest 102:430–437
199. Reguera G, McCarthy KD, Mehta T, Nicoll JS, Tuominen MT, Lovley DR (2005) Extracellular electron transfer via microbial nanowires. Nature 435:1098–1101
200. Richard C, Balavoine F, Schultz P, Ebbesen TW, Mioskowski C (2003) Supramolecular self-assembly of lipid derivatives on carbon nanotubes. Science 300:775–778
201. Rowan B, Wheeler MA, Crooks RM. Patterning bacteria within hyperbranched polymer film templates. Langmuir 2002;18: 9914–7.
202. Sachlos E, Czernuszka JT (2003) Making tissue engineering scaffolds work. Rreview on the application of solid freeform fabrication technology to the production of tissue engineering scaffolds. European Cells and Materials 5:29–40
203. Saggio I, Laufer R (1993) Biotin binders selected from a random peptide library expressed on phage. Biochem J 293:613–616
204. Sanghvi AB, Miller KP-H, Belcher AM, Schmidt CE (2005) Biomaterials functionalization using a novel peptide that selectively binds to a conducting polymer. Nature Mater 4:496–502
205. Sano K, Shiba K (2003) A hexapeptide motif that electrostatically binds to the surface of titanium. J Am Chem Soc 125:14234–14235
206. Sarikaya M, Tamerler C, Jen AK, Schulten K, Baneyx F (2003) Molecular biomimetics: nanotechnology through biology. Nature Mater 2:577–585
207. Sarikaya M, Tamerler C, Schwartz DT, Baneyx F (2004) Materials assembly and formation using engineered polypeptides. Annu Rev Mater Res 34 :373–408
208. Satriano C, Carnazza S, Guglielmino S, Marletta G (2002) Differential cultured fibroblast behavior on plasma and ion beam-modified polysiloxane surfaces. Langmuir 18:9469–9475
209. Satriano C, Carnazza S, Licciardello A, Guglielmino S, Marletta G (2003a) Cell adhesion and spreading on ion-induced micropatterning of polymer surfaces. J Vac Sci Technol A 21:1145–1151
210. Satriano C, Marletta G, Carnazza S, Guglielmino S (2003b) Protein adsorption and fibroblast adhesion on irradiated polysiloxane surfaces. J Mater Sci Mater Med 14:663–670
211. Satriano C., Carnazza S., Guglielmino S., Marletta G (2004) Fibroblast adhesion on polysiloxane surfaces patterned by Ar^+ and He^+ irradiation. 6th International Symposium on Ionizing Radiation and Polymers (IRaP 2004), Houffalize, Belgium
212. Satriano C, Carnazza S, Manso M, Guglielmino S, Rossi F, Marletta G (2005) Enhanced adhesion of U937 cells on irradiated polymer surfaces bio-functionalized by fibronectin adsorption. 19th European Symposium on Biomaterials (ESB 2005), Sorrento, Italy
213. Satriano C, Carnazza S, Guglielmino S, Marletta G (2006a) Bacterial adhesion onto nanopatterned polymer surfaces. Mat Sci Eng C 26: 942–946
214. Satriano C, Marletta G, Guglielmino S, Carnazza S (2006b) Surface free energy and cell adhesion onto irradiated polymer surfaces. In: Mittal KL (ed) Contact angle, wettability and adhesion, 4th edn, pp 1–16
215. Sauer K, Camper AK, Ehrih GD, Costerton JW, Davies DG (2002) *Pseudomonas aeruginosa* displays multiple phenotypes during development as a biofilm. J Bacteriol 184:1140–1154
216. Savage N and Diallo MS (2005) Nanomaterials and water purification: Opportunities and challenges. J Nanoparticle Res 7:331–342

217. Schena M, Yamamoto KR (1988) Mammalian glucocorticoid receptor derivatives enhance transcription in yeast. Science 241:965–967
218. Schindler J, Rovensky L (1994) A model of intrinsic growth of a Bacillus colony. Binary Comput Microbiol 6:105–108
219. Schmidt CE, Shastri VR, Vacanti JP, Langer R (1997) Stimulation of neurite outgrowth using an electrically conducting polymer. Proc Natl Acad Sci USA 94:8948–8953
220. Schwartz MA, Schaller MD, Ginsberg MH (1995) Integrins: Emerging paradigms of signal transduction. Annu Rev Cell Dev Biol 11:549–599
221. Schwartz Z, Lohmann CH, Sisk M, Cochran DL, Sylvia VL, Simpson J, Dean DD, Boyan BD (2001) Local factor production by MG63 osteoblast-like cells in response to surface roughness and 1,25-(OH)2D3 is mediated via protein kinase C- and protein kinase A-dependent pathways. Biomaterials 22:731–741
222. Shalak R, Fox CF (1988) Preface. In: Shalak R, Fox CF (eds) Tissue engineering. Alan R Liss, New York, pp 26–29
223. Sharma, DY, Rao KH (2002) Analysis of different approaches for evaluation of surface free energy of bacterial cells by contact angle goniometry. Adv Coll Interf Sci 98:341–463
224. Shi L, Ardehali R, Caldwell KD, Valint P (2000) Mucin coating on polymeric material surfaces to suppress bacterial adhesion. Colloids Surf B Biointerfaces 17:229–239
225. Shim M, Wong Shi Kam N, Chen RJ, Li Y, Dai H (2002) Functionalization of carbon nanotubes for biocompatibility and biomolecular recognition. Nano Lett 2:285–288
226. Shone C, Wilton-Smith P, Appleton N, Hambleton P, Modi N, Gatley S, Melling J (1985) Monoclonal antibody-based immunoassay for type A *Clostridium botulinum* toxin is comparable to the mouse bioassay. Appl Environ Microbiol 50:63–67
227. Silverman L, Campbell R, Broach JR (1998) New assay technologies for high-throughput screening. Curr Opin Chem Biol 2:397–403
228. Singer II, Scott SD, Kawaka W, Kazazis DM, Gailit J, Ruoslahti E (1988) Cell surface distribution of fibronectin and vitronectin receptors depends on substrate composition and extracellular matrix accumulation. J Cell Biol 106:2171–2182
229. Smith GP, Petrenko VA (1997) Phage display. Chem Rev 97:391–410
230. Soong RK, Bachand GD, Neves HP, Olkhovets AG, Craighead HG, Montemagno CD (2000) Powering an inorganic nanodevice with a biomolecular motor. Science 290: 1555–1558
231. Sorokulova IB, Olsen EV, Chen I, Fiebor B, Barbaree JM, Vodyanoy VJ, Chin BA, Petrenko VA (2005) Landscape phage probes for *Salmonella typhimurium*. J Microbiol Methods 63:55–72
232. Sotiropoulou S, Chaniotakis NA (2003) Carbon nanotube array-based biosensor. Anal Bioanal Chem 375:103–105
233. Souza GR, Christianson DR, Staquicini FI, Ozawa MG, Snyder EY, Sidman RL, Miller JH, Arap W, Pasqualini R (2006) Networks of gold nanoparticles and bacteriophage as biological sensors and cell-targeting agents. Proc Natl Acad Sci USA 103:1215–1220
234. Stenger DA, Gross GW, Keefer EW, Shaffer KM, Andreadis JD, Ma W, Pancrazio JJ (2001) Detection of physiologically active compounds using cell-based biosensors. Trends Biotech 19:304–309
235. Stratmann J, Strommenger B, Stevenson K, Gerlach G-F (2002) Development of a peptide-mediate capture PCR for detection of *Mycobatterium avium* subsp *paratubercolosis* in milk. J Clin Microbiol 40:4244–4250
236. Sulakvelidze A, Alavidze Z, Morris JGJr (2001) Bacteriophage therapy. Antimicrob Agents Chemother 45:649–659
237. Tanaka K, Kashiwagi N, Ogawa T (1988) Effects of light on the electrical output of bioelectrochemical fuel-cells containing *Anabaena variabilis* M-2 mechanisms of the post-illumination burst. J Chem Tech Biotechnol 42:235–240
238. Taton TA, Mirkin CA, Letsinger RL (2000) Scanometric DNA Array Detection with Nanoparticle Probes.Science 289:1757–1760
239. Tirrell M, Kokkoli E, Biesalski M (2002) The role of surface science in bioengineered materials. Sur Sci 500:61–83

240. Toffoli T, Margolus N (1987) In: Cellular Automata Machines. A new environment for modelling. The MIT Press, Cambridge, Massachusetts
241. Tong AH et al (2001) Systematic genetic analysis with ordered arrays of yeast deletion mutants. Science 294:2364–2368
242. Trepel M, Arap W, Pasqualini R (2002) In vivo phage display and vascular heterogeneity: implications for targeted medicine. Curr Opin Chem Biol 6:399–404
243. Triandafillu K, Balazs DJ, Aronsson B-O, Descouts P, Quoc PT, van Delden C, Mathieu HJ, Harms H (2003) Adhesion of *Pseudomonas aeruginosa* strains to untreated and oxygen-plasma treated polyvinyl chloride PVC from endotracheal intubation devices. Biomaterials 24:1507–1518
244. Tsang SC, Davis JJ, Green M, Hill A, Leung YC, Sadler PJ (1995) Immobilization of small proteins on carbon nanotubes: HRTEM study and catalytic activity. J Chem Soc Chem Commun 1803–1804 and 2579
245. Tsang SC, Guo Z, Chen YK, Green M, Hill A, Hambley TW, Sadler PJ (1997) Angew Chem Int Ed Engl 36:2198–2200
246. Tsimring L, Levine H, Aranson I, Ben Jacob E, Cohen I, Shochet O, Reynolds WN (1995) Aggregation patterns in stressed bacteria. Phys Rev Lett 75:1859–1862
247. Tucker CL (2002) High-throughput cell-based assays in yeast. Drug Discov Today 7: S125–S130
248. Turick CE, Tisa LS, Caccavo FJ (2002) Melanin production and use as a soluble electron shuttle for Fe(III) oxide reduction and as a terminal electron acceptor by *Shewanella algae* BrY. Appl Environ Microbiol 68 :2436–2444
249. Ungar F, Geiger B, Ben-Zeev A (1986) Cell contact- and shape-dependent regulation of vinculin synthesis in cultured fibroblasts. Nature 319:787–791
250. Uttamchandani M Mahesh Uttamchandani , Daniel P Walsh , Shao Q Yao , Young-Tae Chang (2005) Small molecule microarrays, recent advances and applications. Curr Opin Chem Biol 9:4–13
251. Valentini RF, Vargo TG, Gardella JAJr., Aebischer P (1992) Electrically conductive polymeric substrates enhance nerve fibre outgrowth in vitro. Biomater 13:183–190
252. van Ginkel CG, Welten HGJ, de Bont JAM, Boerrigter HAM (1986) Removal of ethene to very low concentrations by immobilized *Mycobacterium* E3. J Chem Technol Biotechnol.36:593–598
253. van Loosdrecht MCM, Lyklemam J, Norde W, Zehnder AJB (1990) Hydrophobic and electrostatic parameters in bacterial adhesion. Aquatic Sci 52:103–113
254. van Oss CJ (1989) Energetics of cell-cell and cell-biopolymer interactions. Cell Biophys 14:1–16
255. Vandevivere P, Kirchman DL (1993) Attachment stimulates exopolysaccharide synthesis by a bacterium. Appl Environ Microbiol 59:3280–3286
256. Vidal JC, Garcia E, Castillo JR (1999) In situ preparation of a cholesterol biosensor: entrapment of cholesterol oxidase in an overoxidized polypyrrole film electrodeposited in a flow system: Determination of total cholesterol in serum. Anal Chim Acta 385:213–222
257. Vo-Dinh T, Cullum B (2000) Biosensors and biochip: Advances in biological and medical diagnostics (Review). J Anal Chem 366:540–551
258. Walter G, Bussow K, Cahill D, Lueking A, Lehrach H (2000) Protein arrays for gene expression and molecular interaction screening. Curr Opin Microbiol 3:298–302
259. Wang LF, Yu M (2004) Epitope identification and discovery using phage display libraries: applications in vaccine development and diagnostics. Curr Drug Targets 5:1–15
260. Wang S, Wang H, Onoa GB, Lustig S, Jagota A, Subramoney S, Humphries E, Chung S-Y, Chiang Y-M (2003) Peptides as selective agents for carbon nanotube dispersion. 203rd Meeting on Nanotubes, Nanoscale Materials, and Molecular Devices – Fullerenes, Nanotubes and Carbon Nanostructures – Paris, France, April 27–May 2 2003
261. Warith M, Fernandes L, Gaudet N (1999) Design of in-situ microbial filter for the remediation of naphthalene. Waste Management 19:9–25

262. Wei J, Bagge Ravn D, Gram L, Kingshott P (2003) Stainless steel modified with poly(ethylene glycol) can prevent protein adsorption but not bacterial adhesion. Colloids Surf B Biointerfaces 32:275–291
263. Weiss P (1934) In vitro experiments on the factors determining the course of the outgrowing nerve fiber. J Exp Zool 69:393–448
264. Weiss P (1945) Experiments on cell and axon orientation in vitro: The role of colloidal exudates in tissue organization. J Exp Zool 100:353–386
265. Weiss P (1959) Interactions between cells. Rev Mod Phys 31:449–454
266. Whaley SR, English DS, Hu EL, Barbara PF, Belcher AM (2000) Selection of peptides with semiconductor binding specificity for directed nanocrystal assembly. Nature 405:665–668
267. Willats WG (2002) Phage display: practicalities and prospects. Plant Mol Biol 50:837–854
268. Wojciak-Stothard B, Curtis A, Monaghan W, MacDonald K, Wilkinson C (1996) Guidance and activation of murine macrophages by nanometric scale topography. Exp Cell Res 223:426–435
269. Wolfram S (1984) Cellular Automata as models of complexity. Nature 311:419–424
270. Wong SS, Joselevich E, Woolley AT, Cheung CL, Lieber CM (1998) Covalently functionalized nanotubes as nanometre-sized probes in chemistry and biology. Nature 394:52–55
271. Woo GLY, Mittelman MW, Santerre JP (2000) Synthesis and characterization of a novel biodegradable antimicrobial polymer. Biomaterials 21:1235–1246
272. Wu G, Datar RH, Hansen KM, Thundat T, Cote RJ, Majumdar A (2001) Bioassay of prostate-specific antigen (PSA) using microcantilevers. Nature Biotechnol 19:856–860
273. Yagishita T, Horigome T, Tanaka K (1993) Effects of light, CO_2, and inhibitors on the current output of biofuel cells containing the photosynthetic organism *Synechococcus* sp. J Chem Tech Biotechnol 56:393–399
274. Yamada KM (1997) Integrin signaling. Matrix Biol 16:137–141
275. Yamada KM, Geiger B (1997) Molecular interactions in cell adhesion complexes. Curr Opin Cell boil 9:76–85
276. Yamada KM, Miyamoto S (1995) Integrin transmembrane signaling and cytoskeletal control. Curr Opin Cell Biol 7:681–689
277. Yu J, Smith GP (1996) Affinity maturation of phage-displayed peptide ligands. Methods Enzymol 267:3–27
278. Yue PL, Lowther K (1986) Enzymatic Oxidation of C1 compounds in a Biochemical Fuel Cell. The Chem Eng J 33:B69–B77
279. Zhang X, Halme A (1995) Modeling of a microbial fuel cell process. Biotechnol Let 17: 809–814
280. Zhou J (2003) Microarrays for bacterial detection and microbial community analysis. Curr Opin Microbiol 6:288–294
281. Zhu H, Snyder M (2003) Curr Opin Chem Biol 7:55–63
282. Zhu H et al. (2001) Global analysis of protein activities using proteome chips. Science 293:2101–2105
283. Zuo Y, Maness P-C, Logan BE (2006) Electricity production from steam exploded corn stover biomass. Energy Fuels 20:1716–1721

Light-powered Molecular Devices and Machines

Vincenzo Balzani, Giacomo Bergamini and Paola Ceroni

1 Introduction

In Nature, photons are exploited as energy (e.g., in the photosynthetic process) and elements of information (e.g., in vision). In the last few years, light has also been extensively used in artificial systems at the nanometer scale for energy conversion and information processing [1]. It has been shown that, by assembling suitable molecular components, it is possible to construct molecular-level devices and machines in which light provides the energy needed for performing and/or the signal necessary for monitoring desired functions. The type and utility of the light-related functions depend on the degree of organization of the chemical systems that make use of photons.

2 Bottom-Up Construction of Nanometer Devices and Machines

The idea that atoms could be used to construct nanoscale machines was first raised by R. P. Feynman (*"The principle of physics do not speak against the possibility of maneuvering things atom by atom"*) [2] and depicted in an exciting and visionary way in middle 1980s by K. E. Drexler [3]. However, such an 'atom-by-atom' bottom-up approach to nanotechnology, which seemed so much appealing to physicists [4], did not convince chemists who are well aware of the high reactivity of most atomic species and of the subtle aspects of the chemical bond [5,6].

Vincenzo Balzani
Dipartimento di Chimica "G. Ciamician", Università di Bologna, via Selmi 2,
40126 Bologna, Italy, e-mail: vincenzo.balzani@unibo.it

Giacomo Bergamini
Dipartimento di Chimica "G. Ciamician", Università di Bologna, via Selmi 2,
40126 Bologna, Italy

Paola Ceroni
Dipartimento di Chimica "G. Ciamician", Università di Bologna, via Selmi 2,
40126 Bologna, Italy

S. Bellucci (ed.), *Nanoparticles and Nanodevices in Biological Applications*. Lecture Notes in Nanoscale Science and Technology 7,

Fig. 1 Molecular engineering

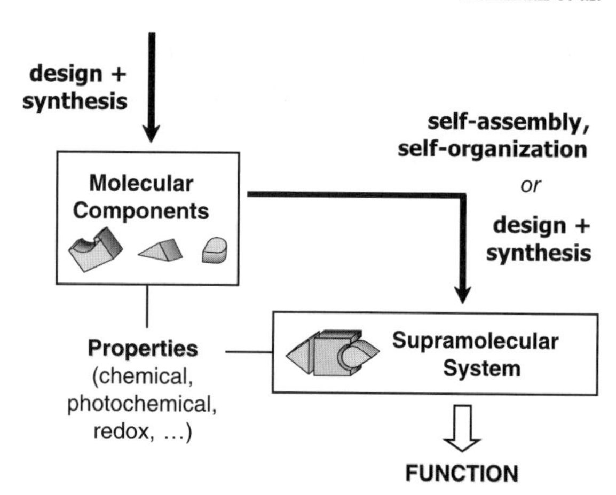

In the late 1970s, in the frame of research on supramolecular chemistry [7], the idea began to arise in the chemical community [8, 9] that molecules are much more convenient building blocks than atoms to construct nanoscale devices and machines. This idea is mainly based on the following points: (i) molecules are stable species, whereas atoms are difficult to handle; (ii) Nature starts from molecules, not from atoms, to construct the great number and variety of nanodevices and nanomachines that sustain life [10]; (iii) most laboratory chemical processes are dealing with molecules, not with atoms; (iv) molecules are objects that already exhibit distinct shapes and carry device-related properties (e.g., properties that can be manipulated by photochemical and electrochemical inputs); (v) molecules can self-assemble or can be connected to make larger structures.

In the following years, supramolecular chemistry grew very rapidly [11] and it became clear that the supramolecular bottom-up approach opens virtually unlimited possibilities (Fig. 1) concerning design and construction of artificial molecular devices and machines [12, 13]. Furthermore, it became more and more evident that such an approach can give invaluable contributions to better understanding the molecular aspects of the extremely complicated nanoscale devices and machines that are responsible for the biological processes [10, 14].

3 Energy Supply

The energy needed for the operation of a molecular device or machine can be supplied in the form of (i) a chemical reagent, (ii) an absorbed photon, or (iii) addition or subtraction of an electron [13]. In view of the shortage of chemical fuels and increasing environmental problems, the ideal primary energy source is sunlight and the worthiest processes are those that do not form waste products. Indeed, even in a knowledge-based society, consumption of non-renewable energy resources and accumulation of waste will continue to pose very difficult problems [15].

In this paper, we will illustrate examples of molecular devices and machines operated by light. Before beginning, it is worthwhile recalling a few basic aspects of the interaction between molecular and supramolecular systems and light. For a more detailed discussion, books [13, 16–18] can be consulted.

4 Molecular and Supramolecular Photochemistry

4.1 Molecular Photochemistry

Figure 2 shows a schematic energy level diagram for a generic molecule that could also be a component of a supramolecular species. In most cases, the ground state of a molecule is a singlet state (S_0), and the excited states are either singlets (S_1, S_2, etc.) or triplets (T_1, T_2, etc). In principle, transitions between states having the same spin value are allowed, whereas those between states of different spin are forbidden. Therefore, the electronic absorption bands observed in the UV-visible spectrum of molecules usually correspond to $S_0 \rightarrow S_n$ transitions. The excited states so obtained are unstable species that decay by rapid first order kinetic processes, namely chemical reactions (e.g. dissociation, isomerization) and/or radiative and non-radiative deactivations. In the discussion that follows, excited state reactions do not need to be explicitly considered and can formally be incorporated within the radiationless decay processes. When a molecule is excited to upper singlet excited states (Fig. 2), it usually undergoes a rapid and 100% efficient radiationless deactivation (internal conversion, *ic*) to the lowest excited singlet, S_1. Such an excited state undergoes deactivation via three competing processes: non-radiative decay to the ground state (internal conversion, rate constant k_{ic}); radiative decay to the ground state (fluorescence, k_{fl}); conversion to the lowest triplet state T_1 (intersystem crossing, k_{isc}). In

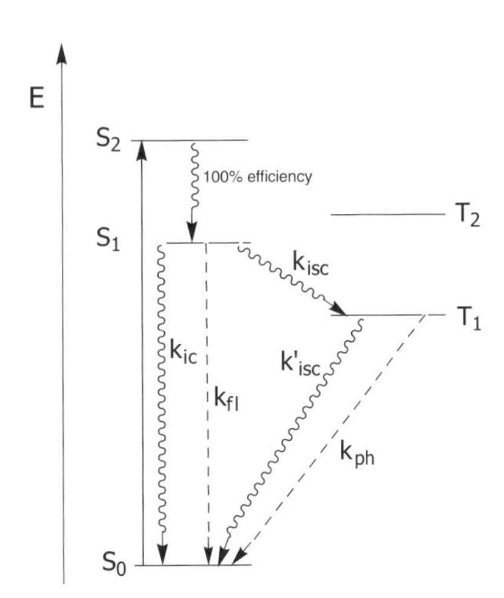

Fig. 2 Schematic energy level diagram for a generic molecule

its turn, T_1 can undergo deactivation via non-radiative (intersystem crossing, k'_{isc}) or radiative (phosphorescence, k_{ph}) decay to the ground state S_0. When the molecule contains heavy atoms, the formally forbidden intersystem crossing and phosphorescence processes become faster. The lifetime (τ) of an excited state, that is the time needed to reduce the excited state concentration by 2.718, is given by the reciprocal of the summation of the deactivation rate constants (Eqs. 1 and 2):

$$\tau(S_1) = \frac{1}{(k_{ic} + k_{fl} + k_{isc})} \tag{1}$$

$$\tau(T_1) = \frac{1}{(k'_{isc} + k_{ph})} \tag{2}$$

The orders of magnitude of $\tau(S_1)$ and $\tau(T_1)$ are approximately $10^{-9} - 10^{-7}$ s and $10^{-3} - 10^{0}$ s, respectively. The quantum yield of fluorescence (ratio between the number of photons emitted by S_1 and the number of absorbed photons) and phosphorescence (ratio between the number of photons emitted by T_1 and the number of absorbed photons) can range between 0 and 1 and are given by Equations 3 and 4:

$$\Phi_{fl} = \frac{k_{fl}}{(k_{ic} + k_{fl} + k_{isc})} \tag{3}$$

$$\Phi_{ph} = \frac{k_{ph} \times k_{isc}}{(k'_{isc} + k_{ph}) \times (k_{ic} + k_{fl} + k_{isc})} \tag{4}$$

Excited state lifetimes and fluorescence and phosphorescence quantum yields of a great number of molecules are known [17].

When the intramolecular deactivation processes are not too fast, that is when the lifetime of the excited state is sufficiently long, an excited molecule $^*A$ in solution may have a chance to encounter a molecule of another solute, B (Eqs. 5–7). In such a case, some specific interaction can occur leading to the deactivation of the excited state by second order kinetic processes. The two most important types of interactions in an encounter are those leading to electron or energy transfer. The occurrence of these processes causes the quenching of the intrinsic properties of $^*A$; energy transfer also leads to sensitization of the excited state properties of the B species. Simple kinetic arguments show that only the excited states that live longer than ca. 10^{-9} s may have a chance to be involved in encounters with other solute molecules.

$$^*A + B \rightarrow A + {}^*B \tag{5}$$

$$^*A + B \rightarrow A^+ + B^- \tag{6}$$

$$^*A + B \rightarrow A^- + B^+ \tag{7}$$

An electronically excited state is a species with quite different properties compared with those of the ground state molecule. In particular, because of its higher energy content, an excited state is both a stronger reductant and a stronger oxidant than the corresponding ground state [18]. To a first approximation, the redox potential of an excited state couple may be calculated from the potential of the related ground state couple and the one-electron potential corresponding to the zero-zero excited state energy, E^{0-0}, as shown by Equations 8 and 9:

$$E(A^+/^*A) \approx E(A^+/A) - E^{0-0} \tag{8}$$

$$E(^*A/A^-) \approx E(A/A^-) - E^{0-0} \tag{9}$$

Detailed discussions of the kinetics aspects of electron- and energy-transfer processes can be found in the literature [19–22].

4.2 Supramolecular Photochemistry

A supramolecular system can be preorganized so as to favor the occurrence of electron- and energy-transfer processes [8, 18]. The molecule that has to be excited, A, can indeed be placed in the supramolecular structure near a suitable molecule, B.

For simplicity, we consider the case of an A–L–B supramolecular system, where A is the light-absorbing molecular unit (Eq. 10), B is the other molecular unit involved with A in the light-induced processes, and L is a connecting unit (often called bridge). In such a system, after light excitation of A, there is no need to wait for a diffusion controlled encounter between *A and B as in molecular photochemistry, since the two reaction partners can already be at an interaction distance suitable for electron and energy transfer:

$$\text{A–L–B} + h\nu \rightarrow {}^*\text{A–L–B} \quad \text{photoexcitation} \tag{10}$$

$$^*\text{A–L–B} \rightarrow \text{A}^+\text{–L–B}^- \quad \text{oxidative electron transfer} \tag{11}$$

$$^*\text{A–L–B} \rightarrow \text{A}^-\text{–L–B}^+ \quad \text{reductive electron transfer} \tag{12}$$

$$^*\text{A–L–B} \rightarrow \text{A–L–}^*\text{B} \quad \text{electronic energy transfer} \tag{13}$$

In the absence of chemical complications (e.g. fast decomposition of the oxidized and/or reduced species), photoinduced electron-transfer processes (Eqs. 11 and 12) are followed by spontaneous back-electron-transfer reactions that regenerate the starting ground state system (Eqs. 11′ and 12′), and photoinduced energy transfer (Eq. 13) is followed by radiative and/or non-radiative deactivation of the excited acceptor (Eq. 13′):

$$\text{A}^+\text{–L–B}^- \rightarrow \quad \text{A–L–B} \quad \text{back oxidative electron transfer} \tag{11′}$$

$$\text{A}^-\text{–L–B}^+ \rightarrow \quad \text{A–L–B} \quad \text{back reductive electron transfer} \tag{12′}$$

$$\text{A–L–}^*\text{B} \rightarrow \quad \text{A–L–B} \quad \text{excited state decay} \tag{13′}$$

In supramolecular systems, electron- and energy-transfer processes are no longer limited by diffusion and occur by first order kinetics. As a consequence, in suitably designed supramolecular systems, these processes can involve even very short-lived excited states.

4.3 $[Ru(bpy)_3]^{2+}$: A Multi-Use Component of Light-Powered Molecular Devices and Machines

Because of a unique combination of chemical stability, redox properties, excited state reactivity, and excited state lifetime, $[Ru(bpy)_3]^{2+}$ (bpy = 2,2′-bipyridine) and related complexes have long been used to obtain photoinduced intermolecular energy and electron transfer processes and have been employed as most valuable components to build up light-powered molecular devices and machines. Our group has been heavily involved in research on complexes of the Ru(II) bipyridine-type family since the early 1970s [23, 24].

On choosing the right counter ion, $[Ru(bpy)_3]^{2+}$ can be dissolved in a variety of solvents, from dichloromethane to water. It is thermodynamically stable and kinetically inert and shows very intense, ligand-centered absorption bands in the UV spectral region and a broad and intense metal-to-ligand-charge-transfer (MLCT) band in the visible region with maximum at 450 nm (Fig. 3). Its lowest excited state, ^{3}MLCT, is reached with unitary efficiency from the upper lying excited states, is relatively long lived (1.1 μs in deaerated acetonitrile solution at 298 K, 5 μs in rigid matrix at 77 K), and exhibits a moderately intense emission around 600 nm ($\Phi = 0.07$ in deaerated acetonitrile at 298 K, Fig. 3). $[Ru(bpy)_3]^{2+}$ has also very interesting electrochemical properties. It shows a metal-centered oxidation process in

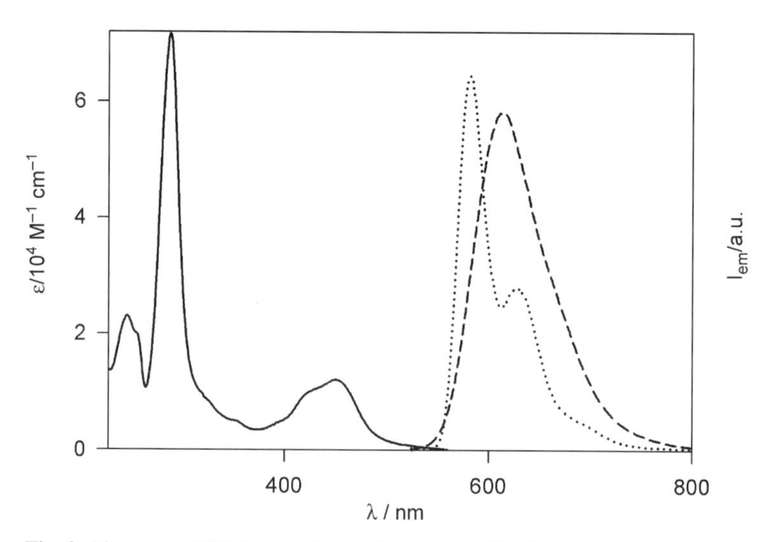

Fig. 3 Absorption (298 K, *solid line*) and emission (298 K, *dashed line*; 77 K, *dotted line*) spectra of $[Ru(bpy)_3]^{2+}$ in acetonitrile solution

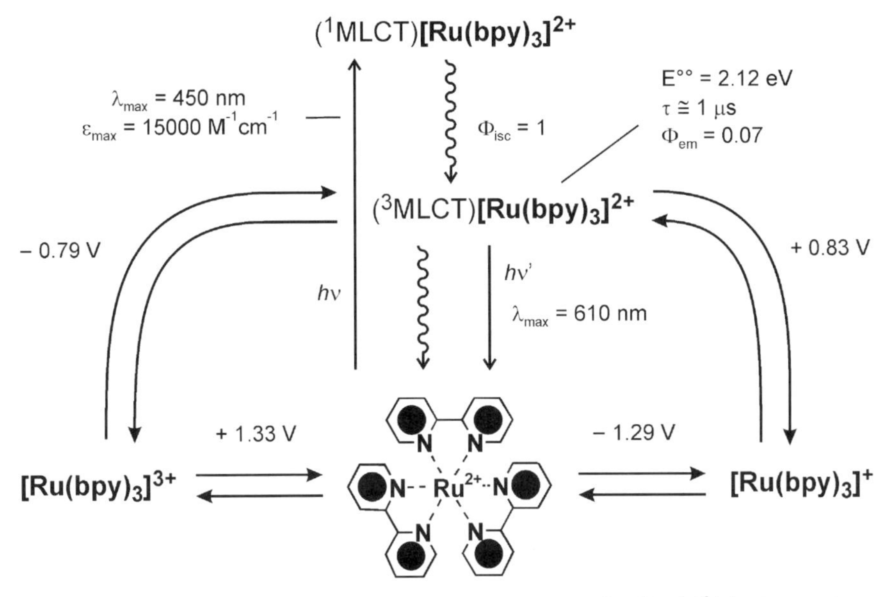

Fig. 4 Schematic representation of some important properties of $[Ru(bpy)_3]^{2+}$ in deaerated acetonitrile solution at 298 K. The potential values are referred to SCE

acetonitrile at room temperature and six distinct ligand-centered reduction processes in dimethylformamide at 219 K [25]. In its ^{3}MLCT excited state, $[Ru(bpy)_3]^{2+}$ is both a good reductant and a good oxidant (Fig. 4). Several hundreds of Ru-polypyridine complexes have been synthesized and characterized since it has been found that the redox and excited state properties can be tuned by changing the ligands or ligand substituents [23, 24].

5 Wires

Most of the systems in which energy transfer has been investigated contain polypyridine metal complexes as donor and acceptor units. Usually, the photoexcited chromophoric group is $[Ru(bpy)_3]^{2+}$ (bpy = 2,2′-bipyridine) and the energy acceptor is an $[Os(bpy)_3]^{2+}$ unit. The excited state of $[Ru(bpy)_3]^{2+}$ playing the role of energy donor is the lowest ^{3}MLCT, which can be obtained by visible light excitation. The occurrence of the energy-transfer process promotes the ground state $[Os(bpy)_3]^{2+}$ acceptor unit to its lowest energy excited state ^{3}MLCT, which lies approximately 0.35 eV below the donor excited state. Both the donor and the acceptor excited states are luminescent, so that the occurrence of energy transfer can be monitored by quenching and/or sensitization experiments with both continuous and pulsed excitation techniques.

Ru(II) and Os(II) polypyridine units have been connected by a variety of bridging ligands and spacers. When the metal-to-metal distance is very short, fast energy transfer occurs by a Förster-type resonance mechanism [13]. In other systems, the two photoactive units are separated by a more or less long spacer. When the spacer is flexible (e.g. $–(CH_2)_n–$ chains), the geometry of the system is not well defined and it is difficult to rationalize the results obtained.

The effect of the nature of the bridge on the rate of energy transfer is clearly demonstrated by the behavior of compounds $\mathbf{1}^{4+}$ [26] and $\mathbf{2}^{4+}$ [27] (Fig. 5) which have almost the same metal-to-metal distance. In compound $\mathbf{1}^{4+}$, in which the two metal-based units are linked by an aromatic bridge, energy transfer occurs with $k > 5 \times 10^{10}\,s^{-1}$ (butyronitrile, 298 K), whereas in compound $\mathbf{2}^{4+}$, in which the bridge is aliphatic, the rate constant is at least three orders of magnitude lower.

The most interesting systems are those in which the two chromophoric units are connected by rigid, modular spacers, as in the case of the $[Ru(bpy)_3]^{2+}–(ph)_n–[Os(bpy)_3]^{2+}$ (ph = 1,4–phenylene; n = 2, 3, 4, 5) species (Fig. 6) [28]. In such compounds, excitation of the $[Ru(bpy)_3]^{2+}$ moiety is followed by energy transfer to the $[Os(bpy)_3]^{2+}$ unit, as shown by the sensitized emission of the latter (CH_3CN, 293 K). The energy-level diagram is schematically shown in Fig. 6. The lowest energy level of the bridge decreases slightly as the number of phenylene units is increased, but always lies above the donor and acceptor levels involved in energy transfer. A further decrease in the energy of the triplet excited state of the spacer would be expected to switch the energy-transfer mechanism from superexchange-mediated to hopping [13]. In the series of compounds shown in Fig. 6, the energy-transfer rate decreases with increasing the length of the oligophenylene spacer (Table 1). Such rate constants are much higher than those expected for a Förster-type mechanism, whereas they can be accounted for by a superexchange Dexter

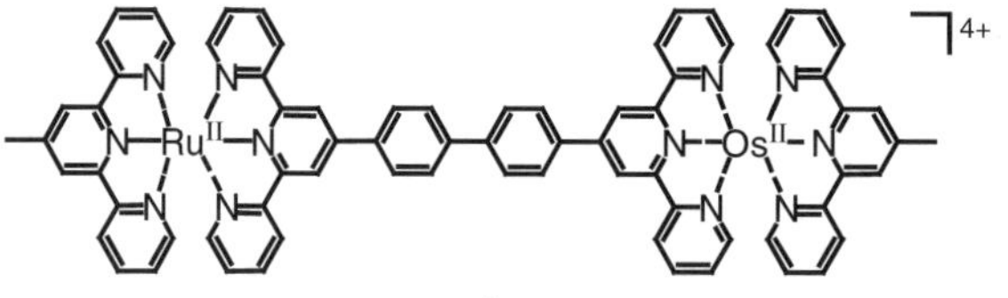

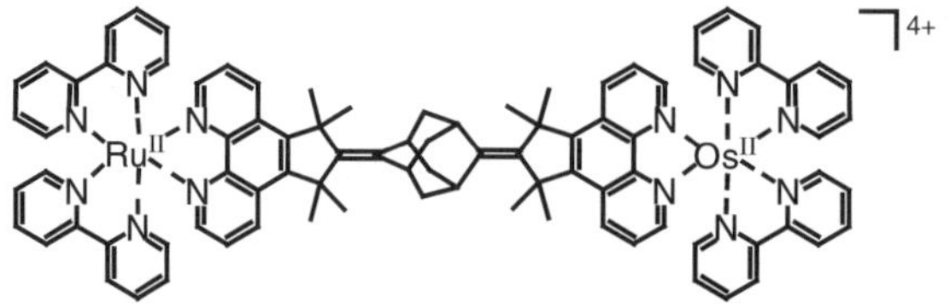

Fig. 5 Binuclear Ru(II)-Os(II) complexes $\mathbf{1}^{4+}$ and $\mathbf{2}^{4+}$ used for investigating the role of the bridge in energy transfer processes [26, 27]

$[Ru(bpy)_3]^{2+}$-$(ph)_n$-$[Os(bpy)_3]^{2+}$ $n = 2, 3, 4, 5$

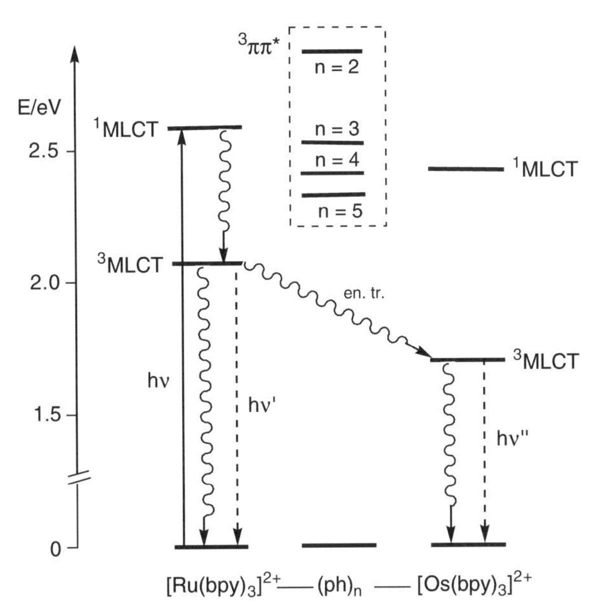

Fig. 6 Structure of compounds $[Ru(bpy)_3]^{2+} - (ph)_n - [Os(bpy)_3]^{2+}$ and energy level diagram for the energy transfer process [28]

mechanism, as suggested by the linear plot obtained for ln k against metal-to-metal distance, with a β value of 0.50 $Å^{-1}$. The values obtained (Table 1) for energy transfer in the series of compounds $[Ru(bpy)_3]^{2+}$–$(ph)_n R_2$–$[Os(bpy)_3]^{2+}$ [29], in which the central phenylene unit carries two hexyl chains, are much lower than those found for the unsubstituted compounds, most likely because the bulky substituents R increase the tilt angle between the phenyl units. A strong decrease in the rate constant is observed when the Ru-donor and Os-acceptor units are linked via an oligophenylene bridge connected in meta position [30].

Quite interesting is the comparison of the above discussed $[Ru(bpy)_3]^{2+}$–$(ph)_n$–$[Os(bpy)_3]^{2+}$ compounds (Fig. 6) with the more recently reported $[Ir(ppyF_2)_2(bpy)]^{+}$–$(ph)_n$–$[Ru(bpy)_3]^{2+}$ (ph = 1,4–phenylene; $n = 2, 3, 4, 5$) systems (Fig. 7) [31]. As reported in Table 1, the Ir-Ru compounds exhibit much higher rate

Table 1 Energy-transfer rate constants for series of donor-bridge-acceptor dinuclear metal complexes [28, 29, 31]

n	$k(s^{-1})$		
	$[Ru(bpy)_3]^{2+}$– $(ph)_n$– $[Os(bpy)_3]^{2+}$	$[Ru(bpy)_3]^{2+}$– $(ph)_nR_2$–$[Os(bpy)_3]^{2+}$	$[Ir(ppyF_2)_2(bpy)]^{+}$– $(ph)_n$–$[Ru(bpy)_3]^{2+}$
2	2.5×10^{11}	–	8.3×10^{11}
3	5.9×10^{10}	6.7×10^{8}	5.9×10^{11}
4	4.1×10^{9}	–	3.6×10^{11}
5	4.9×10^{8}	1.0×10^{7}	3.3×10^{11}
7	–	1.3×10^{6}	–

constants, substantially independent from the length of the spacer. The energy-level diagram of the Ir-Ru complexes, displayed in Fig. 7, shows that the energy level of the donor is almost isoenergetic with the triplet state of the spacers. The energy of the Ir-based donor can, therefore, be transferred to the Ru-based acceptor via the bridging ligand, at least for $n > 2$. This hopping mechanism accounts for the very low dependence of the energy-transfer rate constants on the length of the spacer ($\beta = 0.07$ Å^{-1}).

6 Switches

The electronic properties of a bridging unit can be altered by means of a photonic input. Since, by definition, switching has to be reversible, reversible photochemical reactions have to be used. Photochromic molecules are particularly useful in this regard. An example is given by the D–P–A supramolecular species **3** (Fig. 8) in which photoinduced energy transfer from D to A can be switched by photoexcitation of component P [32]. In such a system, the spacer P is a photochromic fulgide molecule which can be transformed by light in a reversible way between a closed P_a and an open P_b configuration. The donor D is either an anthryl or anthrylvinyl moiety, which can be excited at 258 nm, and the acceptor A is a coumarin molecule. When P is in its closed form P_a (**3a**), its lowest energy level is lower than the energy level of A, so that energy transfer from D to A cannot occur (Fig. 8) and the sensitized luminescence of the coumarin cannot be observed upon excitation of the anthryl moiety. However, when the P species is isomerized with 520-nm light to yield the P_b isomer (**3b**), the energy levels are in scale and the sensitized luminescence of the coumarin component at 500 nm can be observed upon excitation of the anthryl component at 258 nm. Since light of three different wavelengths is needed and four different chromophoric units are involved, such a system is not so easy to handle and its behavior is not really on/off.

An attempt to switch energy transfer process has also been performed with triad $\mathbf{4}^{4+}$ (Fig. 9), which consists of a $[Ru(bpy)]_3^{2+}$ and an $[Os(bpy)]_3^{2+}$ moieties bridged by an anthracene unit [33, 34]. Since the lowest triplet energy level of the anthracene

$[Ir(ppyF_2)_2(bpy)]^+$-$(ph)_n$-$[Ru(bpy)_3]^{2+}$ $n = 2, 3, 4, 5$

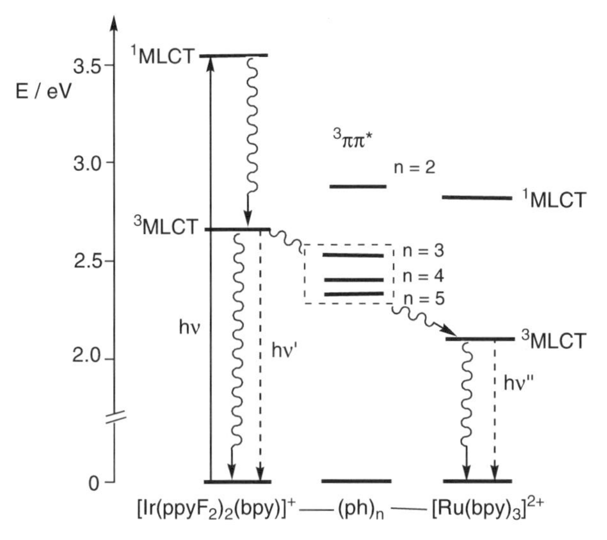

Fig. 7 Structure of compounds $[Ir(ppyF_2)_2(bpy)]^+ - (ph)_n - [Ru(bpy)_3]^{2+}$ and energy level diagram for the energy transfer process [31]

bridge lies in between the lowest triplet MLCT excited state of the Ru- and Os-based complexes, energy transfer from the $[Ru(bpy)]_3^{2+}$ to the $[Os(bpy)]_3^{2+}$ moiety is very efficient. Indeed, continuous irradiation with visible light in deaerated acetonitrile solution causes only the sensitized emission of the Os-based complex. In aerated solution, however, the relatively long-lived excited state of the $[Os(bpy)]_3^{2+}$ moiety sensitizes the formation of singlet oxygen which attacks the anthracene ring to form the endoperoxide derivative $\mathbf{5}^{4+}$ (Fig. 9). As a consequence, the delocalization of the π-system on the bridge is reduced, the lowest energy excited state of the bridge moves to much higher energy, and energy transfer is switched off. In principle, the endoperoxide could be transformed back to anthracene, but such a reaction is difficult to perform. Energy transfer in aerated solution of $\mathbf{4}^{4+}$ has been defined a "self-poisoning" process. The intriguing possibility of designing "self-repairing" processes has been advanced [34].

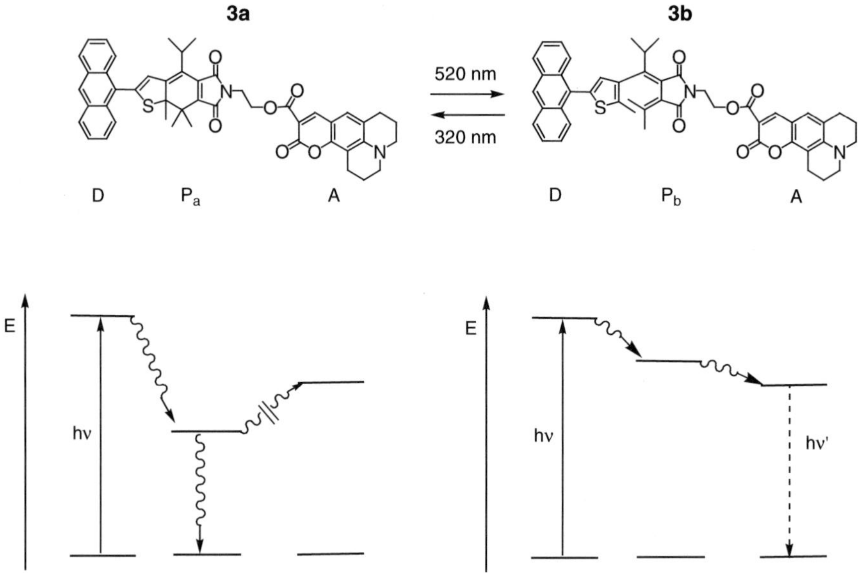

Fig. 8 Switching of energy transfer from an anthracene moiety to a cumarin moiety by photoisomerization of a fulgide bridge [32]

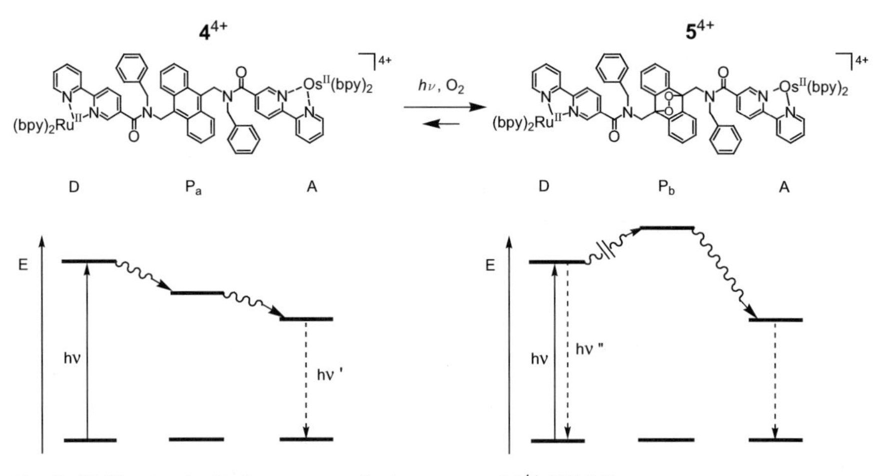

Fig. 9 "Self-poisoning" of energy transfer in compound $\mathbf{4}^{4+}$ [33, 34]

7 Plug-Socket and Extension Cable Systems

Supramolecular species whose components are connected by means of non-covalent forces can be disassembled and reassembled [35] by modulating the interactions that keep the components together, thereby allowing switching of energy- or

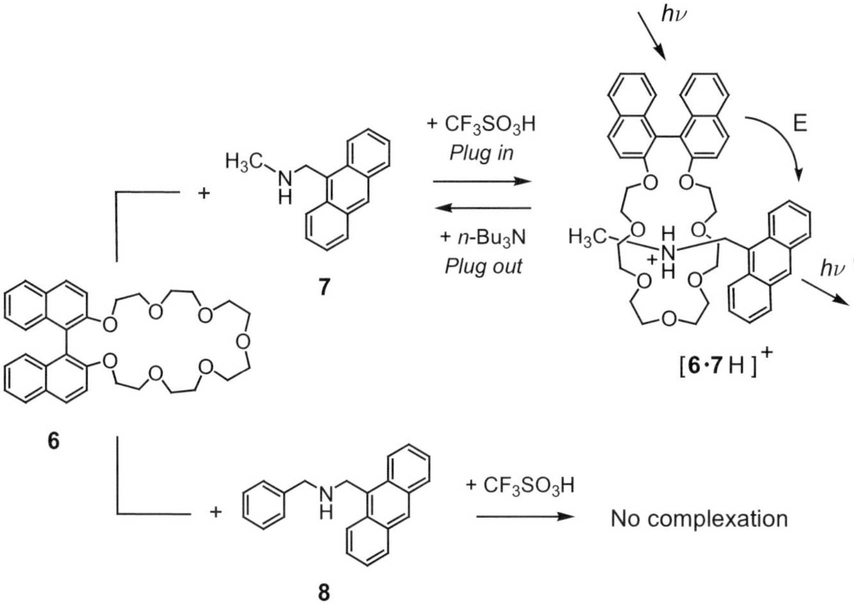

Fig. 10 Switching of photoinduced energy transfer by acid/based controlled plug in/plug out of suitable molecular components [36]

electron-transfer processes. The two-component system shown in Fig. 10 [36] is reminiscent of a plug/socket electrical device and, like its macroscopic counterpart, must be characterized by (i) the possibility of connecting/disconnecting the two components in a reversible way, and (ii) the occurrence of an electronic energy or electron flow from the socket to the plug when the two components are connected (Fig. 10). Hydrogen-bonding interactions between ammonium ions and crown ethers are particularly convenient for constructing molecular-level plug/socket devices since they can be switched on and off quickly and reversibly by means of acid-base inputs. In the system of Fig. 10, the absorption and fluorescence spectra of a CH_2Cl_2 solution containing equal amounts of (±)-binaphthocrown ether **6** and amine **7** indicate the absence of any interaction between the two compounds. However, addition of a stoichiometric amount of acid causes profound changes in the fluorescence behavior of the solution, namely (i) the fluorescence of **6** is quenched, and (ii) the fluorescence of $[7H]^+$ is sensitized upon excitation with light absorbed by **6**. These observations are consistent with the formation of an adduct wherein very efficient energy transfer takes place from the binaphthyl unit of the crown ether to the anthracene group incorporated within the dialkylammonium ion. Such an adduct can be disassembled by the subsequent addition of a stoichiometric amount of base, thereby interrupting the photoinduced energy flow, as indicated by the fact that the initial absorption and fluorescence spectra are restored. Interestingly, the plug-in process does not take place when a plug component incompatible with the size of the socket, such as the benzyl-substituted amine **8**, is employed.

The plug/socket concept can be used to design molecular systems that mimic the function played by a macroscopic electrical extension cable. An extension cable is more complex than a plug/socket device since there are *three* components held together by *two* connections that have to be controllable *reversibly* and *independently*; in the fully connected system, an electron or energy flow must take place between the remote donor and acceptor units.

A system of this type, made of the three components $\mathbf{9}^{2+}$, $\mathbf{10}-\mathrm{H}^{+}$, and $\mathbf{11}^{2+}$, has been reported (Fig. 11) [37]. Component $\mathbf{9}^{2+}$ consists of two moieties: a $[\mathrm{Ru(bpy)_3}]^{2+}$ unit, which behaves as an electron donor under light excitation, and a dibenzo[24]crown-8 macrocycle capable of playing the role of a hydrogen-bonding socket. The extension cable $\mathbf{10}-\mathrm{H}^{+}$ is made up of a dialkylammonium ion, that can insert itself as a plug into a dibenzo[24]crown-8 socket by virtue of hydrogen-bonding interactions, a biphenyl spacer, and a benzonaphtho [36] crown-10 unit, which fulfils the role of a π-electron rich socket. Finally, the $1,1'$-dioctyl-4,4$'$-bipyridinium dication $\mathbf{11}^{2+}$ can play the role of an electron drain plug. In CH_2Cl_2 solution, reversible connection-disconnection of the two plug-socket junctions can be controlled independently by acid-base and red-ox stimulation, respectively, and monitored by changes in the absorption and emission spectra, owing to the different

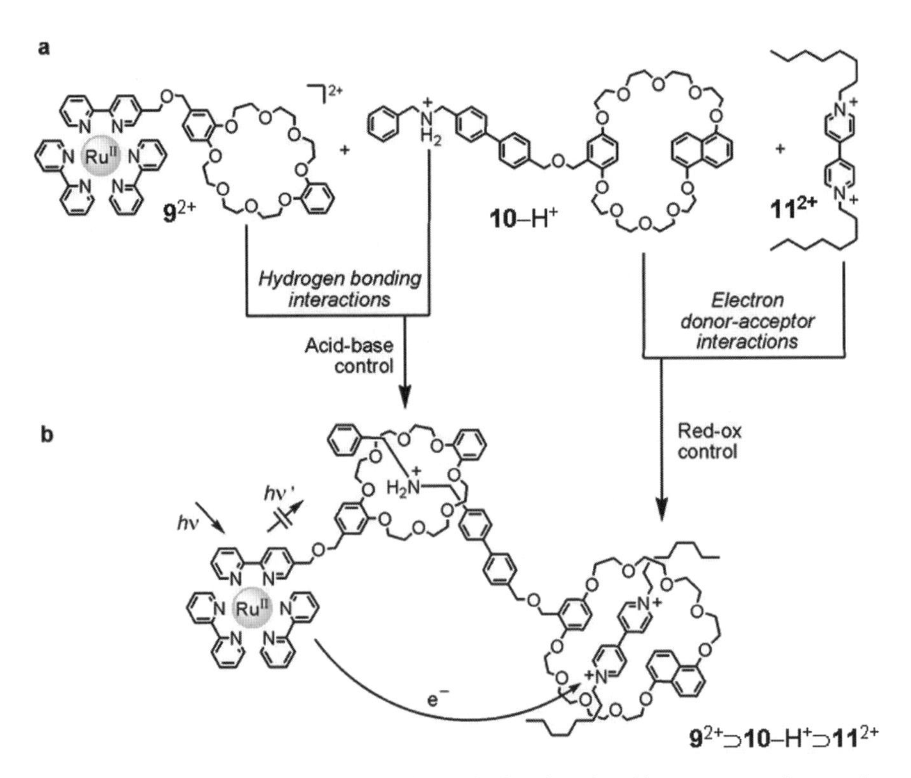

Fig. 11 A supramolecular system which mimics the function played by a macroscopic extension cable [37]

nature of the interactions (hydrogen bonding and π-electron donor-acceptor) that connect the components. In the fully assembled triad, $\mathbf{9}^{2+} \supset \mathbf{10} - H^{+} \supset \mathbf{11}^{2+}$, light excitation of the Ru-based unit of $\mathbf{9}^{2+}$ is followed by electron transfer to $\mathbf{11}^{2+}$, with $\mathbf{10} - H^{+}$ playing the role of an extension cable (Fig. 11). The occurrence of this process is confirmed by nanosecond-laser flash-photolysis experiments, showing a transient absorption signal assigned to the 4,4′-bipyridinium radical cation formed by photoinduced electron transfer within the self-assembled triad. Interestingly, the photoinduced electron-transfer process can be powered by sunlight because the $\mathbf{9}^{2+}$ $[Ru(bpy)_3]^{2+}$-type component shows a broad and intense absorption band in the visible spectral region.

8 Antennas for Light Harvesting

Dendrimers are well defined, tree-like macromolecules, with a high degree of order and the possibility to contain selected chemical units in predetermined sites of their structure [38]. Particularly interesting dendrimers are those containing photoactive components [39]. Because of the close proximity with other units, a photoactive group of a dendrimer can exhibit different properties compared with those exhibited by the same group when it is isolated. For example, in suitably designed dendritic structures, photoexcited units can transfer energy to other components, thereby opening the way towards a number of functions.

In the course of evolution, Nature has succeeded in building up antenna systems that collect an enormous amount of solar energy and redirect it as electronic excitation energy to reaction centres where subsequent conversion into redox chemical energy takes place [40]. Suitably designed dendrimers can mimic the light harvesting function of natural antenna systems since light energy can be channelled by electronic energy transfer towards a specific component of the array [41].

In the dendritic complex $\mathbf{12}^{2+}$ shown in Fig. 12, the 2,2′-bipyridine ligands of the $[Ru(bpy)_3]^{2+}$-type core carry branches containing 1,2-dimethoxybenzene- and 2-naphthyl-type chromophoric units [42]. Since such units (as well as the core) are separated by aliphatic connections, the interchromophoric interactions are weak and the absorption spectra of the dendrimer is substantially equal to the sum of the spectra of the chromophoric groups that are present in its structures. The three types of chromophoric groups, namely, $[Ru(bpy)_3]^{2+}$, dimethoxybenzene, and naphthalene, are potentially luminescent species. In the dendrimer, however, the fluorescence of the dimethoxybenzene- and naphthyl-type units is almost completely quenched in acetonitrile solution, with concomitant sensitization of the luminescence of the $[Ru(bpy)_3]^{2+}$ core ($\lambda_{max} = 610$ nm). These results show that a very efficient energy-transfer process takes place towards the metal-based dendritic core. It should also be noted that in aerated solution, the luminescence intensity of the core is more than twice as intense as that of the $[Ru(bpy)_3]^{2+}$ parent compound because the dendritic branches protect the Ru-bpy based core from dioxygen quenching. Because of the very high absorbance of the naphthyl groups in the UV spectral region, the high

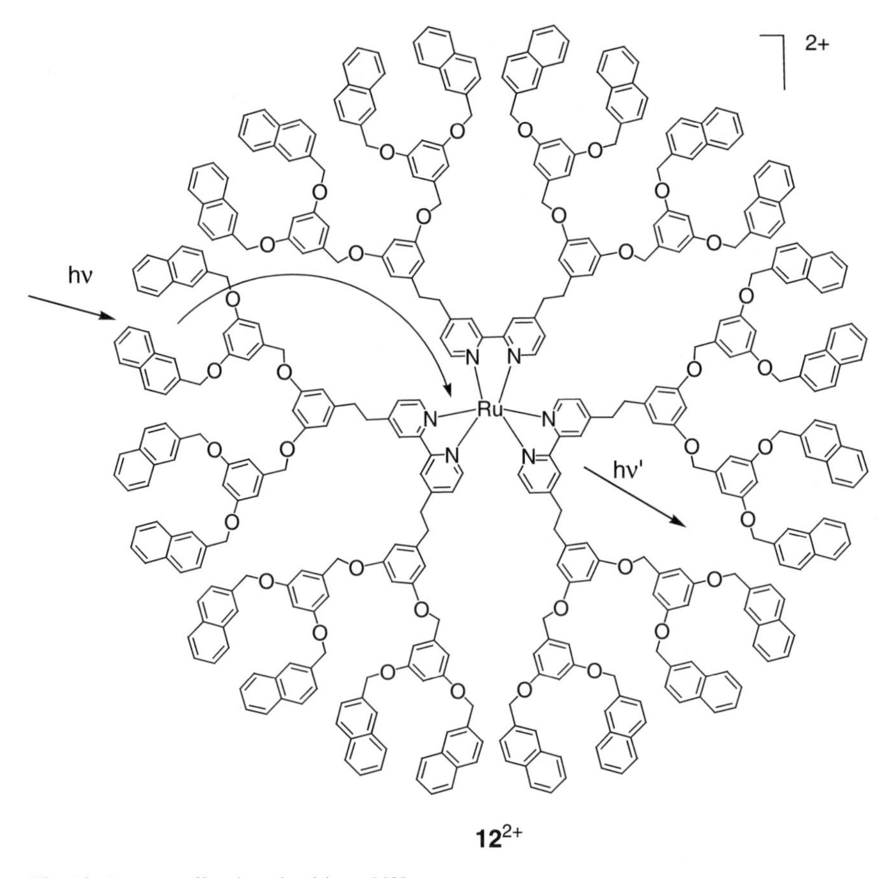

Fig. 12 Antenna effect in a dendrimer [42]

energy-transfer efficiency, and the strong emission of the $[Ru(bpy)_3]^{2+}$-type core, dendrimer $\mathbf{12}^{2+}$ exhibits a strong visible emission upon UV excitation even in very dilute (10^{-7} mol L^{-1}) solutions.

9 Fluorescent Sensors with Signal Amplification

The dendrimers of the poly (propylene amine) family can be easily functionalized in the periphery with luminescent units like dansyl. Each dendrimer nD, where the generation number n goes from 1 to 5, comprises $2^{(n+1)}$ dansyl functions in the periphery and $2^{(n+1)}$-2 tertiary amine units in the interior. Compound **13** (Fig. 13) represents the fourth generation dendrimer 4D containing 30 tertiary amine units and 32 dansyl functions. The dansyl units behave independently from one another so that the dendrimers display light absorption and emission properties characteristic of

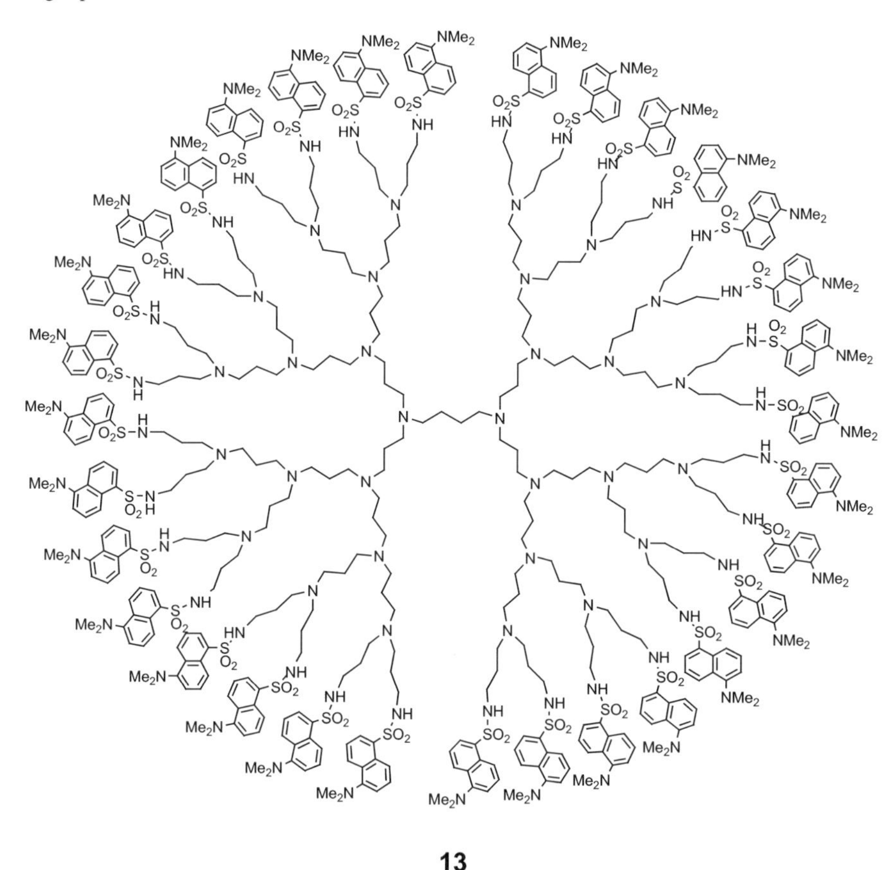

Fig. 13 Fourth generation dendrimer of the poly(propylene amine) family functionalized in the periphery with luminescent dansyl units

dansyl, i.e. intense absorption bands in the near UV spectral region ($\lambda_{max} = 252$ and 339 nm; $\varepsilon_{max} \approx 12000$ and $3900\,\mathrm{L\ mol^{-1}\ cm^{-1}}$, respectively, for each dansyl unit) and a strong fluorescence band in the visible region ($\lambda_{max} = 500\,\mathrm{nm}; \Phi_{em} = 0.46$, $\tau = 16\,\mathrm{ns}$) [44]. Because of the presence of the aliphatic amine in their interior, these dendrimers can play the function of ligands towards transition metal ions.

Coordination of Co^{2+} ions by **13** has been carefully studied [44]. For comparison purposes, the behavior of a monodansyl reference compound has also been investigated. The results obtained have shown that: (i) the absorption and fluorescence spectra of a monodansyl reference compound are not affected by addition of Co^{2+} ions; (ii) in the case of the dendrimer, the absorption spectrum is unaffected, but a strong quenching of the fluorescence of the peripheral dansyl units is observed; (iii) the fluorescence quenching takes place by a static mechanism involving coordination of the metal ion, which is a fully reversible process; (iv) along the series of nD dendrimers, a strong amplification of the fluorescence quenching signal is observed

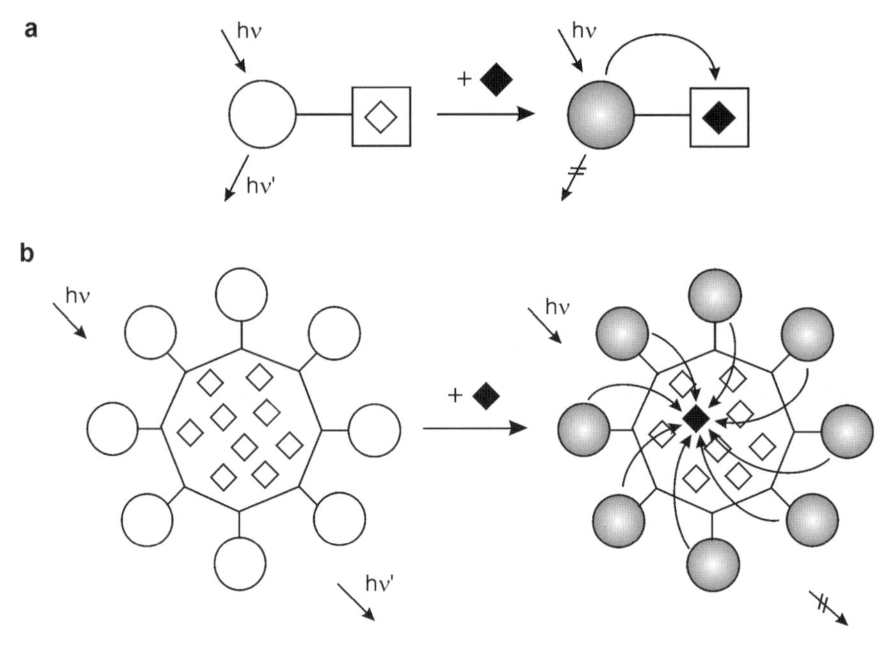

Fig. 14 Schematic representation of (**a**) a conventional fluorescent sensor and (**b**) a fluorescent sensor with signal amplification. Open rhombi indicate coordination sites and black rhombi indicate metal ions. The *curved arrows* represent quenching processes. In the case of a dendrimer, the absorbed photon excites a single fluorophore component that is quenched by the metal ion, regardless of its position [44]

with increasing generation. These results show that dendrimers can be profitably used as supramolecular fluorescent sensors for metal ions. The advantage of a dendrimer for this kind of application is related to the fact that a single analyte can interact with a great number of fluorescent units, which results in signal amplification. For example, when a Co^{2+} ion enters dendrimer **13**, the fluorescence of all the 32 dansyl units is quenched, with a 32 time increase in sensitivity with respect to a normal dansyl sensor. This concept is illustrated in Fig. 14.

10 Logic Gates

In a solid-state transistor, the current flowing from a source to a drain can be modulated by a gate potential. It is possible to design molecular-level photochemionic systems which work on a similar principle, except that the source is a light energy input, the drain is a light energy output (luminescence), and the gate is a chemical input (Fig. 15).

For the sake of space, we will only illustrate two examples of light powered, chemical input(s), optical output logic gates, namely a classical AND gate and a

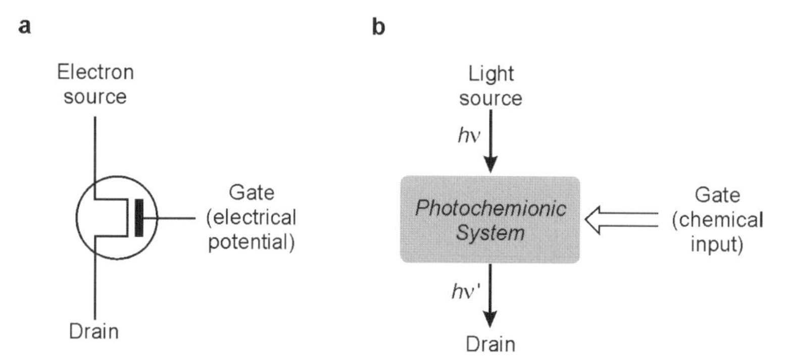

Fig. 15 Schematic illustration of the similarity between a MOSFET electronic transistor (**a**) and a photochemionic gate (**b**)

recently studied system that can perform as XOR and XNOR gates. An exhaustive discussion of molecular level logic gates can be found elsewhere [45].

10.1 AND Logic Gate

The AND operator has two inputs and one output (Fig. 16a) and in a simple electrical scheme, it can be represented by two switches in series. The best examples of molecular level AND gates are those based on two chemical inputs and an optical (fluorescence) output, but examples of molecular systems able to process chemical and optical inputs or two optical inputs with AND functions are also known [45].

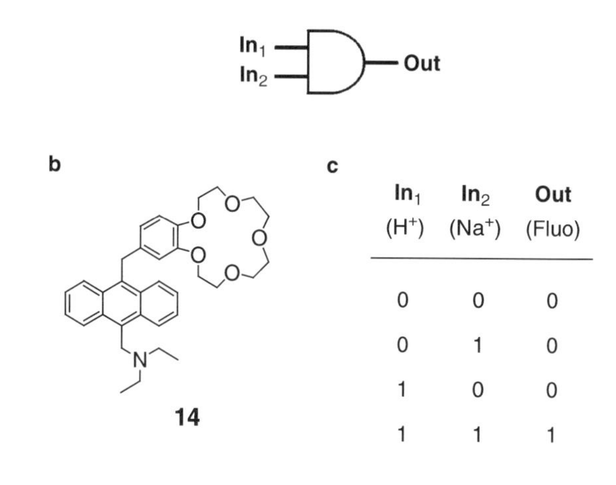

In_1 (H^+)	In_2 (Na^+)	Out (Fluo)
0	0	0
0	1	0
1	0	0
1	1	1

Fig. 16 Symbolic representation (**a**), molecular implementation (**b**), and truth table (**c**) of an AND logic gate based on a three-component system [46]

Figure 16b illustrates the case of a system, **14**, consisting of an anthracene, an aliphatic amine, and a crown ether moieties. The fluorescent excited state of the anthracene component is quenched by electron transfer from the amine and the crown ether components, but such a quenching does not occur when the amine is protonated and the crown ether associates with a Na^+ ion, as indicated in the truth table [46]. In methanol, the fluorescence quantum yield in the presence of 10^{-3} mol $L^{-1}H^+$ and 10^{-2} mol L^{-1} Na^+ is 0.22 (output state 1, fourth line of the truth table, Fig. 16c), whereas none of the three output states 0 has quantum yield higher than 0.009.

10.2 XOR and XNOR Logic Gates

The EXclusive OR (XOR) logic gate is particularly important because it can compare the digital state of two signals. If they are different, an output 1 is given, whereas if they are the same, the output is 0. This logic operation has proven to be difficult to emulate at the molecular scale, but several examples are now available [45,47,48].

1,4,8,11-Tetraazacyclotetradecane (cyclam) in its protonated forms can play the role of host towards cyanide metal complexes. In acetonitrile-dichloromethane 1:1 v/v solution acid-driven adducts formed by $[Ru(bpy)(CN)_4]^{2-}$ with a dendrimer **15** consisting of a cyclam core appended with twelve dimethoxybenzene and sixteen naphthyl units (Fig. 17a). Both $[Ru(bpy)(CN)_4]^{2-}$ and the dendrimer exhibit characteristic absorption and emission bands, in distinct spectral regions, that are strongly affected by addition of acid. When a solution containing equimolar amounts of $[Ru(bpy)(CN)_4]^{2-}$ and **15** is titrated by trifluoroacetic acid, strong spectral changes are observed with isosbestic points maintained up to the addition of two equivalents of acid. The results obtained show that protons promote association of $[Ru(bpy)(CN)_4]^{2-}$ and **15** and that, after addition of two equivalents of acid, a $\{[Ru(bpy)(CN)_4]^{2-} \cdot (2H^+) \cdot \mathbf{15}\}$ adduct is formed, in which the two original species share two protons (Fig. 17a). In the adduct, the fluorescence of the naphthyl units is strongly quenched by very efficient energy transfer to the metal complex, as shown by the sensitized luminescence of the latter [49]. The $\{[Ru(bpy)(CN)_4]^{2-} \cdot (2H^+) \cdot \mathbf{15}\}$ adducts can be disrupted(i) by addition of a base (1,4 diazabicyclo[2.2.2]octane), yielding the starting species $[Ru(bpy)(CN)_4]^{2-}$ and **15**, or (ii) by further addition of triflic acid, with formation of $(\mathbf{15}.2H)^{2+}$ and protonated forms of $[Ru(bpy)(CN)_4]^{2-}$. As a consequence, it has been found that upon stimulation with two chemical inputs (acid and base), $\{[Ru(bpy)(CN)_4]^{2-} \cdot (2H^+) \cdot \mathbf{15}\}$ exhibits two distinct optical outputs (a naphthalene-based (335 nm) and a Ru(bpy)-based (680 nm) emissions) that behave according to an XOR and an XNOR logic, respectively (Fig. 17b).

a

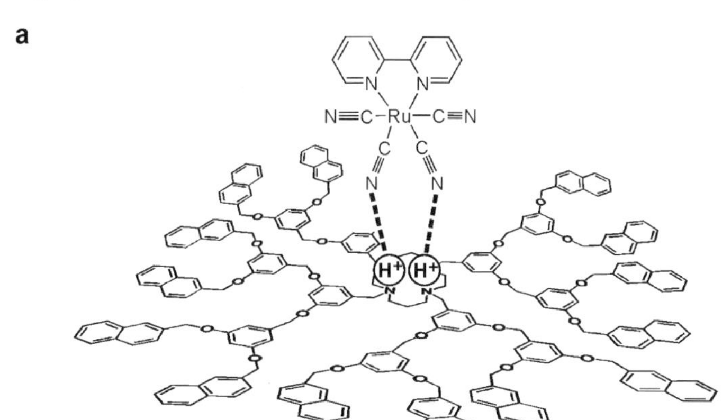

b

IN_1 (acid)	IN_2 (base)	OUT ($I_{335\,nm}$)
0	0	0
0	1	1
1	0	1
1	1	0

XOR

IN_1, IN_2 → OUT

IN_1 (acid)	IN_2 (base)	OUT ($I_{680\,nm}$)
0	0	1
0	1	0
1	0	0
1	1	1

XNOR

IN_1, IN_2 → OUT

Fig. 17 (**a**) Proton driven adduct formation of between $[Ru(bpy)(CN)_4]^{2-}$ and dendrimer **15**. (**b**) Logic behaviour of the $\{[Ru(bpy)(CN)_4]^{2-} \cdot (2H^+) \cdot \mathbf{15}\}$ adduct upon stimulation with acid and base inputs [49]

11 Light Driven Molecular Machines

In green plants, the energy needed to sustain the machinery of life is provided by sunlight. In general, light energy is not used as such to produce mechanical movements, but it is used to produce a chemical fuel, namely ATP, suitable for feeding natural molecular machines [50]. Light energy, however, can directly cause photochemical reactions involving large nuclear movements [8, 18]. A simple example is a photoinduced isomerization from the lower energy *trans* to the higher energy *cis* form of a molecule containing –C=C– or –N=N– double bonds, which is followed by a spontaneous or light-induced back reaction. Such photoisomerization reactions have indeed been used to design molecular machines driven by light energy inputs [51]. In supramolecular species, photoinduced electron-transfer reactions can often cause large displacement of molecular components. Indeed, working with suitable systems, an endless sequence of cyclic molecular-level movements can, in principle, be performed making use of light-energy inputs without generating waste products. Compared to chemical energy inputs, photochemical energy inputs offer other advantages, besides the fundamental one of not generating waste products: (i) Light can be switched on/off easily and rapidly; (ii) Lasers provide the opportunity of working in very small space and very short time domains; (iii) Photons, besides supplying the energy needed to make a machine work, can also be useful to "read" the state of the system and thus to control and monitor the operation of the machine. In the last few years, a great number of light-driven molecular machines have been developed and the field has been extensively reviewed [13]. We will briefly describe a few examples.

11.1 Dethreading/Rethreading of Pseudorotaxanes

Dethreading/rethreading of the wire and ring components of a pseudorotaxane reminds us of the movement of a piston in a cylinder. In order to achieve a light-induced dethreading in such piston/cylinder systems, pseudorotaxane **16** has been designed which incorporates a "light-fueled" motor (i. e., a photosensitiser) in the wire (Fig. 18) [52]. Threading of the wire into the ring is thermodynamically driven because of the electron acceptor and, respectively, electron donor properties of the viologen and crown ether units. In deaerated solution, excitation of the photosensitizer with visible light in the presence of a sacrificial electron donor (e.g., triethanolamine) causes reduction of the electron–acceptor unit and, as a consequence, dethreading takes place. Rethreading can be obtained by allowing oxygen to enter

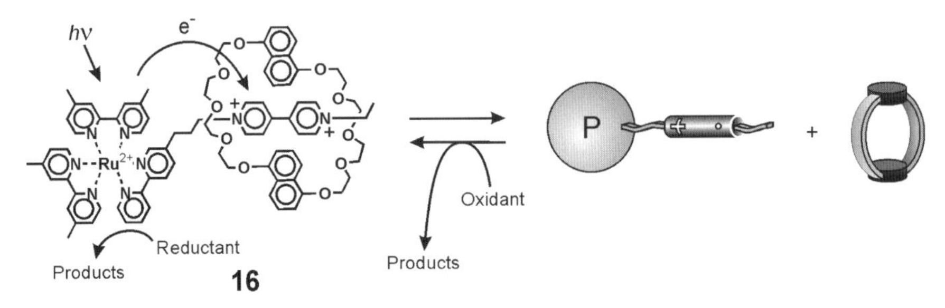

Fig. 18 Light-driven dethreading of pseudorotaxane **16** by excitation of a photosensitizer contained in the wire-type component [52]

the solution. Through a repeated sequence of deoxygenation and irradiation followed by oxygenation, many dethreading/rethreading cycles can be performed on the same solution without any appreciable loss of signal until most of the reductant scavenger is consumed. It should be pointed out, however, that photochemical systems which rely on such a sensitizer-scavenger strategy produce waste species from the decomposition of the reducing scavenger and from the successive consumption of dioxygen.

11.2 A Sunlight Powered Nanomotor

The rotaxane $\mathbf{17}^{6+}$ (Fig. 19) consists of six molecular components suitably chosen and assembled in order to obtain ring shuttling powered by visible light [53]. It comprises a crown ether electron donor macrocycle **R** (hereafter called the ring), and a dumbbell-shaped component which contains two electron acceptor recognition sites for the ring, namely a 4,4′-bipyridinium (A_1) and a 3,3′-dimethyl-4,4′-bipyridinium (A_2) units, that can play the role of "stations" for the ring **R**. Molecular modelling shows that the overall length of $\mathbf{17}^{6+}$ is about 5 nm and the distance between the centers of the two stations, measured along the dumbbell, is about 1.3 nm. Furthermore, the dumbbell-shaped component incorporates a $[Ru(bpy)_3]^{2+}$-type electron transfer photosensitizer **P** which also plays the role of a stopper, a *p*-terphenyl-type rigid spacer **S** which has the task of keeping the photosensitizer far from the electron acceptor units, and finally a tetraarylmethane group **T** as the second stopper. Electrochemical and nuclear magnetic resonance (NMR) spectroscopic data show that the stable conformation of $\mathbf{17}^{6+}$ is by far the one in which the **R** component is located around the better electron acceptor station, A_1^{2+}, as represented in Fig. 19.

The mechanism devised to perform the light-driven shuttling process in the rotaxane $\mathbf{17}^{6+}$ is based on the following four phases:

(a) *Destabilization of the stable conformation*: Excitation with visible light of the photoactive unit **P** (step 1) is followed by the transfer of an electron from the $^*\mathbf{P}$ excited state to the $\mathbf{A_1}$ station, which is encircled by the ring **R** (step 2), with the consequent "deactivation" of this station; such a photoinduced electron-transfer process has to compete with the intrinsic decay of $^*\mathbf{P}$ (step 3).
(b) *Ring displacement*: After reduction ("deactivation") of the $\mathbf{A_1}$ station to $\mathbf{A_1^-}$, the ring moves by Brownian motion to $\mathbf{A_2}$ (step 4), a step that has to compete with back electron-transfer from $\mathbf{A_1^-}$ to the oxidized photoactive unit **P** (step 5). This requirement is the most difficult one to meet since step 4 involves only slightly exergonic nuclear motions whereas step 5 is an exergonic outer-sphere electron transfer process.
(c) *Electronic reset*: A back electron-transfer process from the "free" reduced station $\mathbf{A_1^-}$ to **P** (step 6) restores the electron acceptor power of the $\mathbf{A_1}$ station.
(d) *Nuclear reset*: As a consequence of the electronic reset, the ring moves back again by Brownian motion from $\mathbf{A_2}$ to $\mathbf{A_1}$ (step 7).

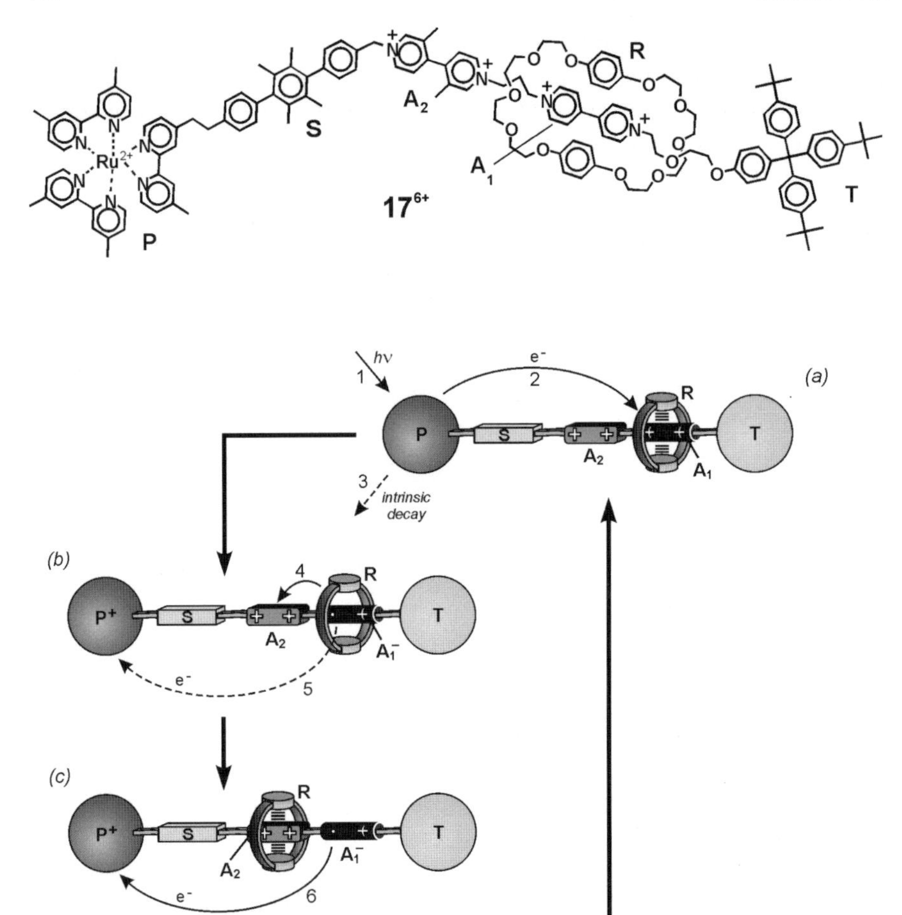

Fig. 19 Structure of rotaxane $\mathbf{17}^{6+}$ and schematic representation of the intramolecular mechanisms for its photoinduced shuttling movement [53]

Reversible displacement of the ring between the two stations $\mathbf{A_1}$ and $\mathbf{A_2}$ should thus be obtained by light energy inputs without consumption of chemical fuels and formation of waste products. It has been found [53] that such photoinduced ring displacement does occur, but with very low efficiency (2%). However, efficiency can be considerably improved (about 12%) in the presence of a suitable external electron relay that, without being consumed, slows down the back electron transfer reaction, thereby leaving more time for ring displacement.

In conclusion, rotaxane **17**$^{6+}$ behaves as an autonomous linear motor powered by visible light. Each phase of the working cycle (Fig. 19) corresponds, in kind, to the fuel injection and combustion (a), piston displacement (b), exhaust removal (c), and piston replacement (d) of a four-stroke macroscopic engine.

The low efficiency of this nanomotor may seem disappointing, but it should be noted that the fuel (sunlight) is free. Besides being powered by sunlight and operating as an autonomous motor, the investigated system shows other quite interesting properties: it works in mild environmental conditions, it is remarkably stable and it can be driven at high frequency (kHz). In principle, when working by a purely intramolecular mechanism, it is also suitable for operation at the single-molecule level.

12 Conclusions

One of the most interesting aspects of supramolecular (multicomponent) systems is their interaction with light. The systems here described show that, in the frame of research on supramolecular photochemistry, the design and construction of nanoscale devices capable of performing useful light-induced functions can indeed be attempted.

The potential applications of photochemical molecular devices and machines are various – from energy conversion to sensing and catalysis – and, to a large extent, still unpredictable. As research in the area is progressing, two interesting kinds of nonconventional applications of these systems begin to emerge: (i) their behavior can be exploited for processing information at the molecular level [54] and, in the long run, for the construction of chemical computers [55]; (ii) their mechanical features can be utilized for transportation of nanoobjects, mechanical gating of molecular-level channels, and nanorobotics [56].

However, it should be noted that the species described here, as most multicomponent systems developed so far, operate in solution, that is, in an incoherent fashion and without control of spatial positioning. Although the solution studies are of fundamental importance to understand their operation mechanisms and for some use (e.g. drug delivery), it seems reasonable that before such systems can find applications in many fields of technology, they have to be interfaced with the macroscopic world by ordering them in some way. The next generation of multicomponent molecular species will need to be organized so that they can behave coherently and can be addressed in space. Viable possibilities include deposition on surfaces, incorporation into polymers, organization at interfaces, or immobilization into membranes or porous materials. Recent achievements in this direction [57–60] let one optimistically hope that useful devices based on functional (supra)molecular systems could be obtained in a not too distant future.

Apart from foreseeable applications related to the development of nanotechnology, investigations on photochemical molecular devices and machines are important to increase the basic understanding of photoinduced reactions and other processes such as self-assembly, as well as to develop reliable theoretical models. This research has also the important merit of stimulating the ingenuity of chemists, thereby instilling new life into Chemistry as a scientific discipline.

Acknowledgment Financial support from MIUR (PRIN "Progettazione e caratterizzazione di dispositivi molecolari aftificiali").

References

1. *Handbook of Photochemistry and Photobiology*, Vols. 1–4, ed by H.S. Nalwa (American Scientific Publishers, Stevenson Ranch 2003)
2. (a) R.P. Feynman: Eng. Sci. **23**, 22 (1960); (b) R.P. Feynman: Saturday Rev. **43**, 45 (1960)
3. K.E. Drexler: *Engines of Creation – The Coming Era of Nanotechnology* (Anchor Press, New York 1986)
4. K.E. Drexler: *Nanosystems. Molecular Machinery, Manufacturing, and Computation* (Wiley, New York 1992)
5. R.E. Smalley: Sci. Am. **285**, 68 (2001)
6. G. M. Whitesides: Sci. Am. **285**, 70 (2001)
7. J.-M. Lehn: Angew. Chem. Int. Ed. Engl. **27**, 89 (1988)
8. V. Balzani, L. Moggi, F. Scandola, In: *Supramolecular Photochemistry* ed by V. Balzani (Reidel, Dordrecth 1987) p 1
9. J.-M. Lehn: Angew. Chem. Int. Ed. Engl. **29**, 1304 (1990)
10. D.S. Goodsell: *Bionanotechnology – Lessons from Nature* (Wiley, New York 2004)
11. *Encyclopedia of Supramolecular Chemistry*, ed by J.L. Atwood, J.W. Steed (Dekker, New York 2004)
12. V. Balzani, A. Credi, M. Venturi: Chem. Eur. J. **8**, 5524 (2002)
13. V. Balzani, A. Credi, M. Venturi: *Molecular Devices and Machines – A Journey in the Nano World* (Wiley-VCH, Weinheim 2003)
14. F. Cramer: *Chaos and Order. The Complex Structure of Living Systems* (VCH, Weinheim 1993)
15. N. Armaroli, V. Balzani: Angew. Chem. Int. Ed. Engl. **45**, 2 (2006)
16. A. Gilbert, J. Baggott: *Essentials of Molecular Photochemistry* (Blackwell Science, London 1991)
17. *Handbook of Photochemistry,* 3rd edn, ed by M. Montalti, A. Credi, L. Prodi, M.T. Gandolfi, (CCR, Taylor and Francis, New York 2006)
18. V. Balzani, F. Scandola: *Supramolecular Photochemistry* (Horwood, Chichester 1991)
19. *Electron Transfer in Chemistry*, Vols. 1–5, ed by V. Balzani (Wiley-VCH, Weinheim 2001)
20. *Adv. Chem. Phys.*, special volumes 106–107, ed by M. Bixon, J. Jortner In: Electron Transfer: From Isolated Molecules to Biomolecules (1999)
21. V. May, O. Kühn: *Charge and Energy Transfer Dynamics in Molecular Systems* (Wiley-VCH, Weinheim 2000)
22. H.B. Gray, J.R. Winkler: Q. Rev. Biophys. **36**, 341 (2003)
23. V. Balzani, A. Juris: Coord. Chem. Rev. **211**, 97 (2001)
24. A. Juris, V. Balzani, F. Barigelletti, S. Campagna, P. Belser, A. von Zelewsky: Coord. Chem. Rev. **84**, 85 (1988)
25. S. Roffia, R. Casadei, F. Paolucci, C. Paradisi, C. A. Bignozzi, F. Standola: J. Electroanal. Chem. 302 157 (1991)
26. F. Barigelletti, L. Flamigni, V. Balzani, J.-P. Collin, J.-P. Sauvage, A. Sour, E.C. Constable, A.M.W. Cargill Thompson: J. Chem. Soc. Chem. Commun. **942**, (1993)

27. V. Balzani, F. Barigelletti, P. Belser, S. Bernhard, L. De Cola, L. Flamini: J. Phys. Chem. **100**, 16786 (1996)
28. S. Welter, N. Salluce, P. Belser, M. Groeneveld, L. De Cola: Coord. Chem. Rev. **249**, 1360 (2005)
29. B. Schlicke, P. Belser, L. De Cola, E. Sabbioni, V.Balzani: J. Am. Chem. Soc. **121**, 4207 (1999)
30. A. D'Aleo, S. Welter, E. Cecchetto, L. De Cola: Pure Appl. Chem.**77**, 1035 (2005)
31. S. Welter, F. Lafolet, E. Cecchetto, F. Vergeer, L. De Cola: ChemPhysChem **6**, 2417 (2005)
32. J. Walz, K. Ulrich, H. Port, H.C. Wolf, J. Wonner, F. Effenberger: Chem. Phys. Lett. **213**, 321 (1993)
33. P. Belser, R. Dux, M. Baak, L. De Cola, V. Balzani: Angew. Chem. Int. Ed. Engl. **34**, 595 (1995),
34. L. De Cola, V. Balzani, P. Belser, R. Dux, M. Baak: Supramol. Chem. **5**, 297 (1995)
35. V. Balzani, A. Credi, M. Venturi: Proc. Natl. Acad. Sci. **99**, 4814 (2002)
36. E. Ishow, A. Credi, V. Balzani, F. Spadola, L. Mandolini: Chem. Eur. J. **5**, 984 (1999)
37. B. Ferrer, G. Rogez, A. Credi, R. Ballardini, M.T. Gandolfi, V. Balzani, Y. Liu, H.-R. Tseng, J.F. Stoddart: Proc. Natl. Acad. Sci. **103**, 18411 (2006)
38. (a) G.R. Newkome, F. Vögtle: *Dendrimers and Dendrons* (Wiley-VCH, Weinheim 2001) (b) *Dendrimers and Other Dendritic Polymers*, ed by J.M.J. Fréchet, D.A. Tomalia (Wiley, New York 2001)
39. See e.g.: (a) P. Ceroni, G. Bergamini, F. Marchioni, V. Balzani: Prog. Polym. Sci. **30**, 453(2005) (b) F.C. De Schryver, T. Vosch, M. Cotlet, M. Van der Auweraer, K. Müllen, J. Hofkens: Acc. Chem. Res. **38**, 514 (2005)
40. T. Pullerits, V. Sundström: Acc. Chem. Res. **29**, 381 (1996)
41. V. Balzani, P. Ceroni, M. Maestri, V. Vicinelli: Curr. Opinion Chem. Biol. **7**, 657 (2003)
42. M. Plevoets, F. Vögtle, L. De Cola, V. Balzani: New. J. Chem. 23, **63** (1999)
43. F. Vögtle, S. Gestermann, C. Kauffmann, P. Ceroni, V. Vicinelli, L. De Cola, V. Balzani: J. Am. Chem. Soc. **121**, 12161 (1999)
44. (a) V. Balzani, P. Ceroni, S. Gestermann, C. Kauffmann, M. Gorka, F. Vögtle: Chem. Commun. **853**, (2000) (b) V. Balzani, P. Ceroni, V. Vicinelli, S. Gestermann, M. Gorka, C. Kauffmann, F. Vögtle: J. Am. Chem. Soc. **122**, 10398 (2000)
45. (a) A.P. de Silva, H.Q.N. Gunaratne, T. Gunnlaugsson, A.J.M. Huxley, C.P. McCoy, J.T. Rademacher, T.E. Rice: Chem. Rev. **97**, 1515 (1997) (b) V. Balzani, A. Credi, M. Venturi: ChemPhysChem **3**, 101 (2003)
46. A.P. de Silva, H.Q.N. Gunaratne, C.P. McCoy: J. Am. Chem. Soc. **119**, 7891 (1997)
47. A. Credi, V. Balzani, S.J. Langford, J.F. Stoddart: J. Am. Chem. Soc. **119**, 2679 (1997)
48. F. Pina, M.J. Melo, M. Maestri, P. Passaniti, V. Balzani, J. Am. Chem. Soc. **122**, 4496 (2000)
49. G. Bergamini, C. Saudan, P. Ceroni, M. Maestri, V. Balzani, M. Gorka, S.-K. Lee, J. van Heyst, F. Vögtle: J. Am. Chem. Soc. **126**, 16466 (2004)
50. D.-P. Hader, M. Tevini: *General Photobiology* (Pergamon, Oxford 1987)
51. See, e.g.: (a) S. Shinkai, T. Nakaji, T. Ogawa, K. Shigematsu, O. Manabe: J. Am. Chem. Soc. **103**, 111 (1981) (b) N. Koumura, R.W.J. Zijlstra, R.A. van Delden, N. Harada, B.L. Feringa: Nature **401**, 152 (1999)
52. P.R. Ashton, V. Balzani, O. Kocian, L. Prodi, N. Spencer, J.F. Stoddart: J. Am. Chem. Soc. **120**, 11190 (1998)
53. V. Balzani, M. Clemente-León, A. Credi, B. Ferrer, M. Venturi, A. H. Flood, J. F. Stoddart: Proc. Natl. Acad. Sci. **103**, 1178 (2006)
54. For a representative recent example, see: D. Margulies, C.E. Felder, G. Melman, A. Shanzer: J. Am. Chem. Soc. **129**, 347 (2007)
55. (a) D. Rouvray: Chem. Brit. **36**, 46 (2000) (b) P. Ball: Nature **406**, 118 (2000)
56. A.A.G. Requicha: Proc. IEEE **91**, 1922 (2003)
57. J. Berná, D.A. Leigh, M. Lubomska, S.M. Mendoza, E.M. Pérez, P. Rudolf, G. Teobaldi, F. Zerbetto: Nature Mat. **4**, 704 (2005)
58. A. Kocer, M. Walko, W. Meijberg, B.L. Feringa: Science **309**, 755 (2005)

59. A.P. de Silva, M.R. James, B.O.F. Mckinney, D.A. Pears, S.M. Weir: Nature Mat. **5**, 787 (2006)
60. S. Bhosale, A.L. Sisson, P. Talukdar, A. Furstenberg, N. Banerji, E. Vauthey, G. Bollot, J. Mareda, C. Roger, F. Würthner, N. Sakai, S. Matile Science **313**, 84 (2006)

Hofmeister Effects in Enzymatic Activity, Colloid Stability and pH Measurements: Ion-Dependent Specificity of Intermolecular Forces

Andrea Salis, Maura Monduzzi and Barry W. Ninham

1 Introduction

Electrolytes are integral components of biological systems, and they are involved in several enzymatic pathways essential to life. However, the high degree of specificity of electrolytes in determining mechanisms of enzymatic action is at best only partially understood.

That situation, the nature and origin of specific ion, or Hofmeister effects, is universal in physical chemistry [1].

In elementary physical chemistry, a distinction is made between those two kinds of electrolytes. Conventionally, strong electrolytes are those fully dissociated in water. Weak electrolytes are only partially dissociated in water. The distinction then assigns to weak electrolytes a very important role since they modify the acid/base equilibria of water solutions. The simultaneous presence of an electrolyte in water solution in both its undissociated and dissociated forms gives rise to a pH buffer. The equilibrium between the two forms opposes pH variations caused by the addition of strong acids or bases to the system. Strong (fully dissociated) electrolytes were believed to have little effect on pH. This is approximately true only at low ionic strength [2], as it will be shown in Sect. 4.

Enzymes, as for all proteins, contain a large number of acidic and basic groups located mainly on the exterior "surface". When placed in aqueous media, the superficial net charge of the enzyme can change as a result of bulk pH modifications that affect the acid/basic dissociation equilibria. Consequently enzymatic activity,

Andrea Salis
Dipartimento di Scienze Chimiche, Università di Cagliari – CSGI, Cittadella Monserrato, S.S. 554 Bivio Sestu, 09042 Monserrato, Italy

Maura Monduzzi
Dipartimento di Scienze Chimiche, Università di Cagliari – CSGI, Cittadella Monserrato, S.S. 554 Bivio Sestu, 09042 Monserrato, Italy

Barry W. Ninham
Dipartimggeo di Scienze Chimiche, Università di Cagliari – CSGI, Cittadella Monserrato, S.S. 554 Bivio Sestu, 09042 Monserrato, Italy Department of Applied Mathematics, A.N.U. Canberra, Australia

S. Bellucci (ed.), *Nanoparticles and Nanodevices in Biological Applications*. Lecture Notes in Nanoscale Science and Technology 7,

structural features and solvation, can change radically. Changes in bulk pH may affect charge distribution on the substrate and product also. In effect, charge variations at the interfacial enzymatic "surface" will be reflected in changes in the binding of the substrate, and the catalytic efficiency. It is commonly held that the effect of pH on the rate of an enzymatic reaction can be explained by assuming that only one of the possible charged enzymatic forms gives rise to the optimal catalytic performance. In other words, there is an optimum pH value that favors the maximum concentration of the enzyme-substrate intermediate.

Ionic strength is another variable parameter that may affect catalytic activity. At high ionic strength, a lowering of the carboxylic acid pK_as may occur (probably due to competition of the cation with hydronium for the carboxylic site), but at neutral pH, little effect on the overall charge of the enzyme molecule is generally observed, unless the variation in charge occurs within the active site.

In the last decade, it has become clear that such first order theoretical notions are too crude to characterize real enzyme performance. Although ionic strength and pH are recognized as important factors that affect enzyme activity, specific ion effects are not embraced by classical (electrostatic) theories of physical chemistry. Indeed enzyme conformations, stability and activity are a result of a complex interplay of factors other than electrostatic forces alone. These are usually subsumed under separate competing terms like electrostatic, dipolar, and van der Waals interactions, hydrogen bonds, solvation and polarization effects, association–dissociation equilibria of charged groups. In reality, these terms, which have their origin in theories of interactions between molecules in dilute media, are mnemonics for a much richer class of many body forces that operate in solution, for which the distinctions between them are smeared.

The story starts more than a century ago (see Fig. 1)

The different effects of neutral salts on the solubility of proteins were first explored systematically by Franz Hofmeister in 1888 [3]. His historical papers have been recently translated into English and republished [4].

For proteins, the precipitation (salting out) / solubilization (salting in) efficiency of the anions, at a fixed ionic strength, was found to decrease/increase in the order:

$$HPO_4^{2-} > SO_4^{2-} > F^- > Cl^- > Br^- > NO_3^- > I^- > ClO_4^- > SCN^-$$

The phenomenon embraced by this sequence is referred as the Hofmeister series (HS), or alternatively as a lyotropic series. It is represented schematically in Fig. 2. From the beginning, a puzzling recurrent observation was the salting out efficiency followed more or less in the HS order when the pH is higher than the isoelectric point (pI) and in the opposite order at $pH < pI$ [5–8]. This was explained only recently [9].

Enzymatic activity also seemed to follow Hofmeister series as reported in several works from 1960 to the present time [10–22]. Sometimes, the series reversed with change of buffer [14].

Sometimes, the specific activating/deactivating effect of ions was recognized as a Hofmeister effect. Sometimes, it was then being attributed to other factors [23]. The effects are well known to biochemists. Nevertheless, the development of a theoret-

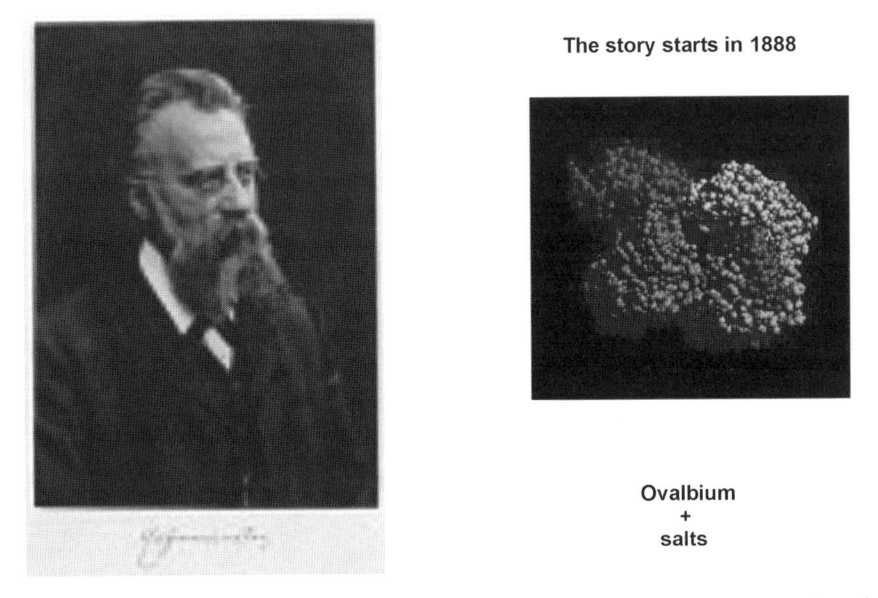

Fig. 1 Franz Hofmeister (1850–1922). In 1901 Hofmeister proposed that life results from the activity of enzymes, and that a specific enzyme is responsible for every vital reaction [3]

ical framework that allows a description of the ion-specific interplay between ionic strength, activity coefficients, pH, and enzyme activity remained elusive.

"Specific ion" effects occur not just in biochemistry, but almost everywhere in physical, colloid, polymer and surface chemistry [1, 24–30].

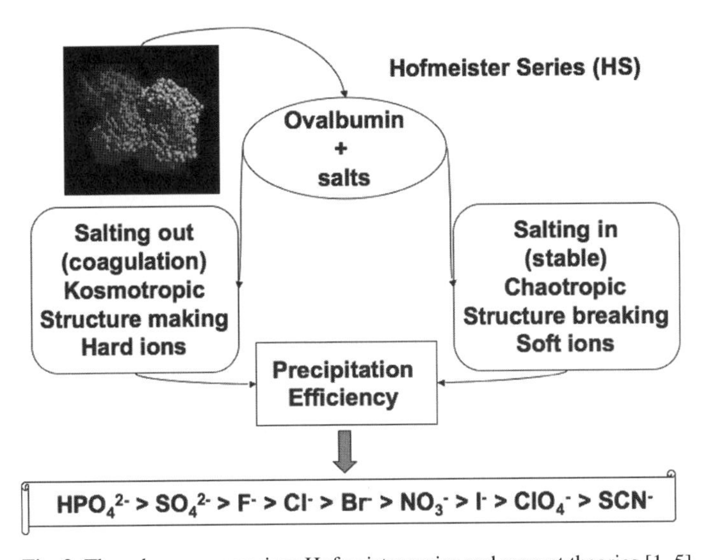

Fig. 2 The scheme summarizes Hofmeister series and current theories [1–5]

Until recently, they have remained inexplicable with conventional theories of solution and colloid chemistry. But some progress is now under way [1].

Hofmeister himself remained bemused at the source of his phenomena, whether they were due to bulk or surface (adsorption) effects. Both involve hydration effects induced by ions [31–34]. Hofmeister lent towards attributing the matter principally to changes in bulk water structure. From this point of view, the ion specificity is determined by the ability of ions to form (kosmotropic), or to break (chaotropic) hydrogen bonds in water systems. This is the standard view of the matter [32–35].

But a host of morphological changes due to ion specific headgroup area variations in surfactant systems and in direct force measurements implicate surface effects [36]. The same is true for a number of other phenomena discussed below. It seems that in general, both bulk and surface (hydration) effects must be involved. A more recent approach stems from the rather surprising recognition that the theoretical treatment of nonelectrostatic forces experienced by ions near an interface must be treated at the same level as the forces due to the classical electrostatic potential [37]. These nonelectrostatic, electrodynamic fluctuation (called for brevity, dispersion or NES) forces are treated only in linear approximation in the standard framework. This topic will be resumed below in Sect. 3 after the presentation of some experimental results related to enzyme activity in the following Sect. 2. It will be demonstrated that both bulk and surface effects play a role [9,38]. In this context, it is also worth mentioning that, according to very recent results, ions affect the first hydration shells only. Ions neither enhance nor weaken the hydrogen bond network, at least over the time scale experienced by femtosecond pump spectroscopy [39,40]. Thus, Hofmeister effects are likely to find explanation in interactions of the ions with the macromolecule and its first hydration shell [41].

2 Enzymatic Activity: Role of Buffers, Salts, pH and Ionic Strength

2.1 The Case of the Lipase from Aspergillus niger

The enzymatic activity of lipase from *Aspergillus niger* was first measured as a function of pH. To do this, the ratio $[H_2PO_4^-]/[HPO_4^{2-}]$ was changed to prepare different sodium phosphate buffer solutions in the range of pH 5.0–7.5 ($[H_2PO_4^-] + [HPO_4^{2-}] = 5\,mM$), and without any added salts. The results are shown in Fig. 3a. Lipase A shows a maximum activity at pH = 6, that is $A_s = 160\,\mu mol\,min^{-1}\,g^{-1}$.

Then, some experiments were performed in the presence of added sodium salts, using a phosphate buffer at initial pH=6. Activity data and pH data are shown in Fig. 3b. While pH values decrease with increasing salt concentrations as expected, different trends in the enzymatic activity are observed for the different anions. Independent of the salt induced pH decrease, it is remarkable that a high concentration of the most chaotropic anions, namely SCN^- and ClO_4^-, causes a decrease of activity

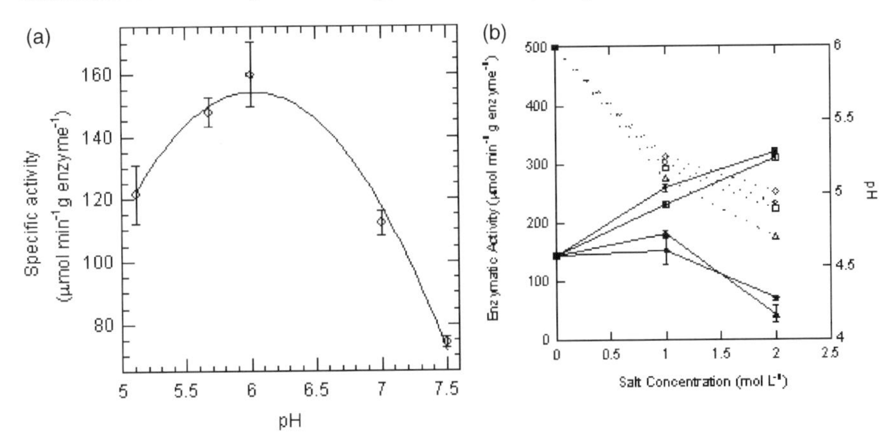

Fig. 3 Enzymatic activity of lipase from *Aspergillus niger* in phosphate buffer 5 mM solution at different pH values (**a**), and in the presence of different sodium salts (Br^-, Cl^-, ClO_4^-, SCN^-) (**b**). The effect of salts on pH (*dashed lines* in the graph on the right side) is also shown

below the value measured in the buffer at initial pH=6. On the contrary, NaCl and NaBr induce similar superactivities. These trends should be compared to those observed in the phosphate buffer at initial pH=7 where all salts were found to increase enzyme activity in the order $Br^- > Cl^- \approx NO_3^- > ClO_4^-$.

Fig. 4 shows the data obtained using the buffer at initial pH = 7. Strikingly, enzymatic activity increased significantly, even though the addition of NaBr induced a further significant decrease of the (measured) pH below the optimum value of 6 (Fig. 3a). These results emphasize the crucial role of Br^- in increasing the enzymatic activity. The addition of Bromide salts having different monova-

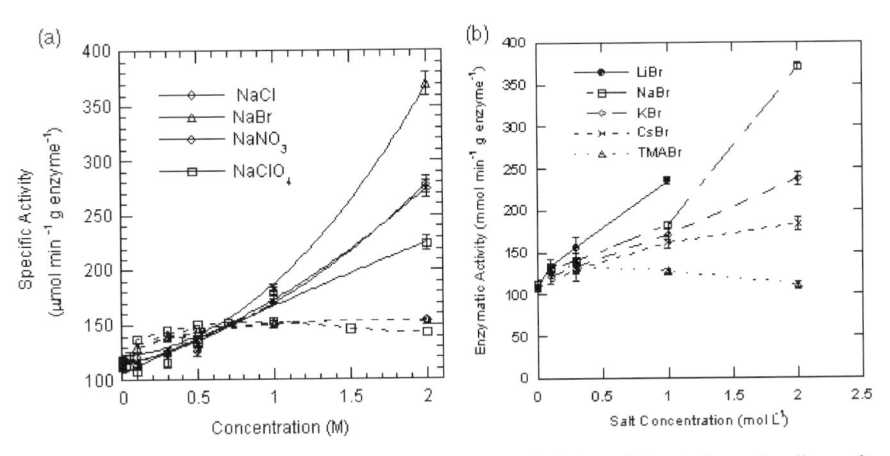

Fig. 4 Specific ion Effects on *Aspergillus niger* lipase hydrolytic activity. Anions of sodium salts (**a**), and of the cations of bromide salts (**b**). In the left side graph, the solid lines show the experimental activities, the dashed lines the calculated activities

lent cations (see Fig. 4b) confirmed the ionic pair Na^{+} Br^{-} as a specific salt able to promote the superactivity of the lipase A from *Aspergillus Niger* (from 160 to $370\,\mu mol\,min^{-1}\,g^{-1}$).

But this is valid in the presence of a phosphate buffer. Indeed, other buffers produce different trends of the enzymatic activity as a function of pH values. This is shown in Fig. 5. This is likely to suggest different stories.

From this work, some important insights into an understanding of the nature of the Hofmeister effects seem to emerge. If the major effect of salt addition was due to the salt induced water structure modification (bulk phenomena), the corresponding pH change should have produced the increase of enzymatic activity predicted by theoretical curves (dashed lines in Fig. 4a). But this is not so. The present results can be justified only in terms of a specific interaction (adsorption) of Br^{-} anions at the enzyme surface. To these interactions, ion specific dispersion forces not part of the present intuition on these matters are likely to contribute. The final outcome is that the enzyme is driven, by these additional adsorption forces, with consequent changes in protein hydration, to a conformation very active for the catalytic process.

Hence, the main conclusion is that enzymatic activity is intimately related to anion specific surface phenomena. In addition, it should here be recalled that surface adsorption phenomena will occur at the glass electrode surface also (see below Sect. 4). Consequently, the measured pH may not be the real bulk pH of the solution. One may not even be sure of the meaning of the pH reading. But the superactivity induced by adding specifically the ion pair NaBr 2M in phosphate buffer at initial pH = 7 (which is not the optimal pH without added salts) seems to be a quite remark-

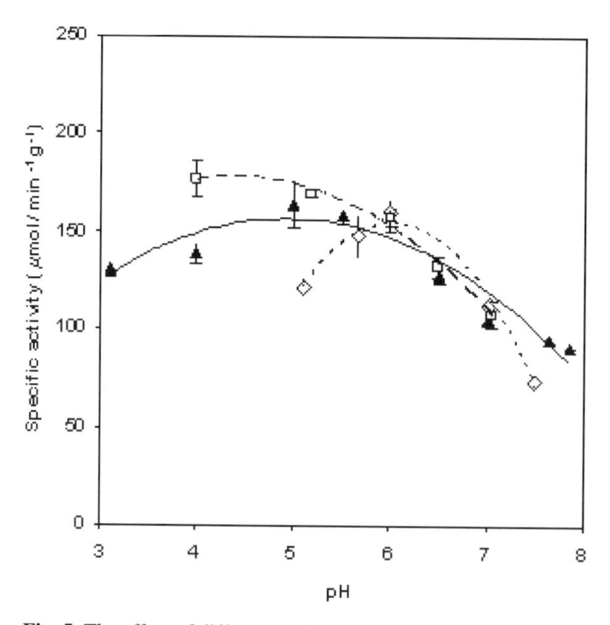

Fig. 5 The effect of different buffers on the enzymatic activity of the lipase from *Aspergillus niger*. (▲) Sodium citrate; (◇) Sodium phosphate; (□) Tris-HCl

able result. In summary, it turns out that it is more convenient to obtain the apparent optimal pH conditions by adding salts rather than by using an "ad hoc" buffer solution only. The phenomenon may well be quite general, in which case, it provides a clue to specific ion regulation mechanisms and superactivities of enzymes.

2.2 The Case of the Lipase from Candida rugosa

The effects of weak and strong electrolytes on the enzymatic activity of *Candida rugosa* lipase have also been explored. Weak electrolytes, used as buffers, set pH, while strong electrolytes regulate ionic strength. The interplay between pH and ionic strength has been assumed to be the determinant of enzymatic activity. In experiments that probe activities by varying these parameters, there has been little attention focussed on the role of specific electrolyte effects. Here, for *Candida rugosa* lipase, is another specific case where both buffers, and the choice of background electrolyte ion pair, strongly affect the enzymatic activity. The effects are dramatic at high salt concentration. Indeed, a 2M concentration of NaSCN is able to fully inactivate the lipase. By contrast, Na_2SO_4 acts generally as an activator, whereas NaCl shows a quasi-neutral behavior. Such specific ion effects are well known and are classified among "Hofmeister effects". But there has been little awareness of them, or of their potential for optimization of activities in the enzyme community. Rather than the effects per se, the focus here is on their origin. New insights into mechanism can be reasonably proposed as a result of the new observations.

Fig. 6a shows *Candida rugosa* lipase activity as a function of pH obtained with three different buffers, namely sodium citrate, sodium phosphate and Tris-HCl, at 5mM concentration. The main result is that the activity depends on both the pH and the specific weak electrolyte used to prepare the buffer. From this fact, two other points emerge:

The pH for maximum activity varies with the buffer. It is about 5.4 for Tris-HCl, 6 for citrate, and 6.7 for phosphate;

The maximum activity of *Candida rugosa* lipase has different values for each buffer, i.e. phosphate $\sim$ citrate $>$ Tris-HCl.

Fig. 6b shows the experiments performed by using more concentrated buffers (200 mM). Comparing with the previous situation, the curve maxima are closer, falling in a very narrow pH range (6.5–6.6). Moreover, the enzymatic activity decreases significantly in the order phosphate $>$ citrate $>$ Tris-HCl.

These results are likely to hold specifically for *Candida rugosa* lipase only. Indeed, similar measurements performed in 5mM buffers using the lipase from *Aspergillus niger* (see Fig. 5) also showed a slight specific dependence on the buffer anion/cation. However, unlike the results above, the highest activity was observed with Tris-HCl buffer as shown in Fig. 5. This is a strong indication that specific adsorption of buffer ions at substrate and enzyme is involved.

The activity of the *Candida rugosa* lipase, in the presence of both weak and strong electrolytes, was then assayed. Different salts (concentrations: 0.5 and 1M

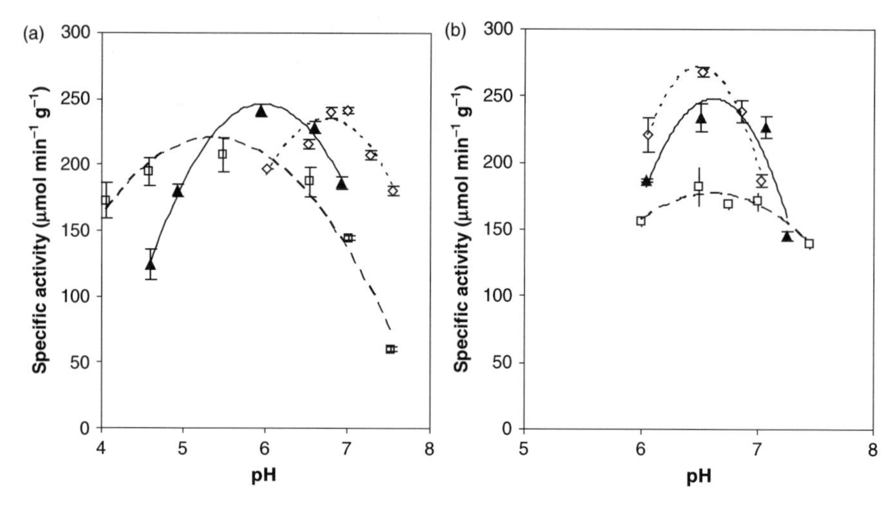

Fig. 6 The effect of different buffers and of their concentration [(**a**) 5 mM, (**b**) 200 mM] on the maximum activity of the lipase from *Candida rugosa*. (▲) Sodium citrate; (◇) Sodium phosphate; (□) Tris-HCl

of Na_2SO_4, 0.5, 1 and 2M of NaCl and NaSCN) together with the substrate p-nitrophenyl acetate were dissolved in different buffer solutions, namely phosphate (Ph) and Tris-HCl, (5 and 200 m M) at the initial pH = 7. If we use the standard nomenclature of Hofmeister series, Na_2SO_4 is considered to be kosmotropic, NaCl is neutral and NaSCN is chaotropic. Prior to each enzymatic activity determination, pH measurements were performed by means of a glass electrode.

The substrate solutions in the presence of both buffers and salts were then used for the determination of the enzymatic activity. Results are reported in Fig. 7 for Phosphate buffer and Fig. 8 for Tris-HCl buffer, respectively. In all cases, the three salts act in a similar way, in particular, the sulfate is generally activating, chloride is neutral or slightly activating/deactivating, and SCN^- is strongly deactivating. Only in the presence of Tris-HCl 5 mM (Fig. 8) is the sulfate slightly deactivating.

2.2.1 Problems with Ion Specificity

Ion specific effects on enzymatic activity have been a matter of debate for more than forty years [10, 11]. Theoretical models have not been predictive, even at a qualitative level. One reason for that situation is now clear. The entire theoretical framework used, even for pH, was flawed in its omission of a class of forces, dispersion forces, largely implicated in producing the effects. The problem is complicated by the superposition of several different effects difficult to disentangle. But the fact that these Hofmeister effects can reverse order, with enzyme, with buffer, above and below the point of zero charge, with interchange of cation, etc. [9], all point to the diversity of competitive adsorption available to dispersion forces acting on ions in concert with varying adsorbate.

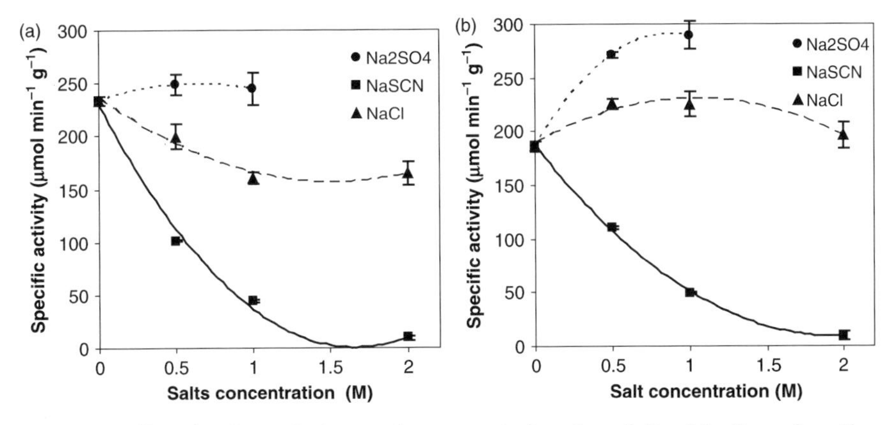

Fig. 7 The effect of sodium salts (type and concentration) on the activity of the lipase from *Candida rugosa* in the presence of phosphate buffer 5 mM (**a**) and 200 mM (**b**)

The purpose of this discussion is to rehearse these matters in detail to pinpoint where the problems lie. Once identified, these will be discussed separately. Then, we try to rationalize them at a qualitative level taking into account contributions from previously neglected nonelectrostatic dispersion forces. The issues are the following:

- Both pH and salt addition affect enzyme activity. These effects are usually treated separately via simple kinetic models (see below Sect. 2.2.2). The second is usually discussed only in terms of ionic strength variations. No ion specificity is usually taken into account. A recent work explained Hofmeister effects with the bulk pH variations induced by the salt addition [42]. This cannot be done at high

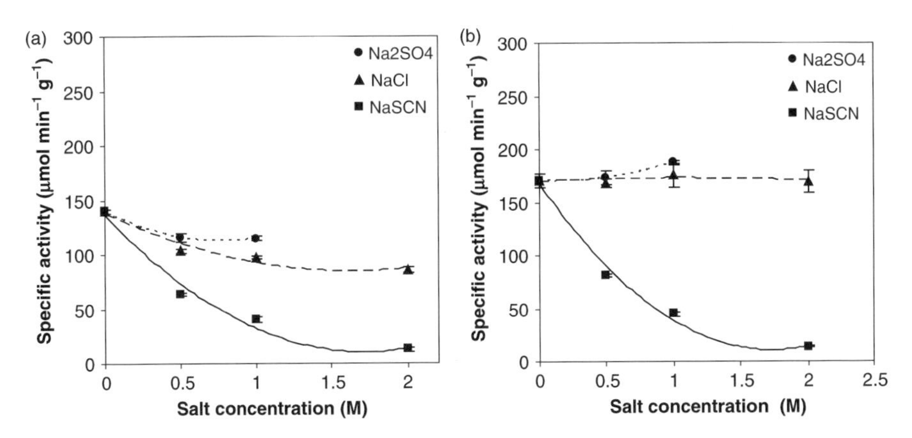

Fig. 8 The effect of sodium salts (type and concentration) on the activity of the lipase from *Candida rugosa* in the presence of Tris-HCl buffer 5 mM (**a**) and 200 mM (**b**)

salt concentration as shown above in the case concerning the enzymatic activities of *Aspergillus niger* lipase [19].

- When high concentrations of salts are added to a buffer/protein solution, the bulk pH is apparently modified. With proteins, the pH variation follows a direct or a reverse Hofmeister series depending on the pI [9]. Dispersion forces have been already used to rationalize these effects (see Sect. 2.2.3) [43]. This point is further complicated by the fact that the most commonly used technique for pH measurements, a pH-meter equipped with a glass electrode, is likely to be affected by the presence of high concentrations of salts (see below Sect. 4). Direct or reverse Hofmeister series can be obtained depending on the buffer [38]. The result is that the measured pH likely differs from the real pH of the solution. The origin of this artifact is still unquantified since a full explanation of the phenomenon has not yet been given (see Sect. 2.2.3). It is known, however, that the standard theory behind interpretation of pH, a purely electrostatic theory, ignores specific ion adsorption and interactions.
- On the basis of the structure of *Candida rugosa* lipase (see Sect. 2.2.4 and Fig. 10), new insights into the effects caused by salt addition and pH changes on enzymatic activity will be proposed. pH and salt addition affect amino acid residues both at the active site and on the exterior "surface" of the enzyme. A different charge of the catalytically involved amino acids and a structure distortion may occur. The balance between these two effects is likely to be responsible for the different activity trends measured with different enzymes (see Sect. 2.2.5).

2.2.2 Failure of Conventional Kinetic Models to Explain the Effect of pH and Ionic Strength on Enzymatic Activity

Enzymes are affected by pH. This is due to a combination of factors:

- The binding of the substrate to the enzyme.
- The catalytic activity of the enzyme.
- The degree of ionization of the substrate.
- The variation of the structure of the enzyme (usually assumed important only at extreme pH values) [44].

A simple model that captures enzyme activity dependence on pH is the following:

$$
\begin{array}{ccccc}
E^- & & ES^- & & \\
K_{E2}\ \Big\updownarrow\ H^+ & & K_{ES2}\ \Big\updownarrow\ H^+ & & \\
EH + S & \underset{k_{-1}}{\overset{k_1}{\rightleftharpoons}} & ESH & \xrightarrow{k_2} & EH + P \\
K_{E1}\ \Big\updownarrow\ H^+ & & K_{ES1}\ \Big\updownarrow\ H^+ & & \\
EH_2^+ & & ESH_2^+ & &
\end{array}
$$

where: K_{E1}, K_{E2}, K_{ES1} and K_{ES2} are the ionization constants of the ionizable amino acids (i.e. His and Glu) in the active site.

From this model, a Michaelis/Menten-type relationship can be derived [44]:

$$v = \frac{V'_{MAX}[S]}{K'_M[S]} \tag{1}$$

where

$$V'_{MAX} = \frac{V_{MAX}}{\frac{[H^+]}{K_{ES1}} + 1 + \frac{K_{ES2}}{[H^+]}} \tag{2}$$

$$K'_M = K_M \frac{\frac{[H^+]}{K_{E1}} + 1 + \frac{K_{E2}}{[H^+]}}{\frac{[H^+]}{K_{ES1}} + 1 + \frac{K_{ES2}}{[H^+]}} \tag{3}$$

Kinetic parameters K_M and V_{MAX} are pH dependent. Nothing is supposed to be affected by the buffer used to obtain the desired pH value. Our results in Figs. 5–6 clearly show the inadequacy of this model.

It has also been reported that ionic strength is an important parameter affecting enzyme activity [45]. If the reaction rate depends upon the approach of charged moieties – as for the rate controlling step in the catalytic mechanism of chymotrypsin that involves the approach of two positively charged groups (His57 and Arg145) – increasing the ionic strength of the solution causes a significant increase/decrease in k_{cat} [45]. The approximate relationship below, that is a consequence of classical electrolyte theory, has been proposed:

$$\log k = \log k_0 + Z_A Z_B I^{1/2} \tag{4}$$

Here, k is the measured rate constant, k_0 is the zero ionic strength rate constant, Z_A and Z_B are the electrostatic charges of the reacting species, and I is the ionic strength of the solution. From this equation, we can see that the reaction rate decreases when the charges are opposite in sign. It increases when the charges are identical. Results in Figs. 7–8 show that the addition of monovalent salts NaCl and NaSCN, which should cause the same ionic strength effect, leads instead to a very different enzymatic activity from that predicted.

2.2.3 Effect of Salt Addition on pH of Proteins and Buffers Solutions: Dispersion Forces

It is an easy experiment to verify that the measured pH of a protein/buffer solution changes with the addition of strong electrolytes. Measured pH is ion specific and follows a Hofmeister series.

Some recent papers by Boström et al. [9, 43, 46] predicted the behavior of some proteins in water solution in the presence of salts by taking into account both electrostatic and dispersion forces at the same level. They used the Poisson–Boltzmann cell model to calculate the net protein charge, surface pH, solution pH, and ion distributions.

The net protein charge, surface pH, solution pH, and ion distributions can be determined self-consistently via the non-linear Poisson–Boltzmann equation for monovalent ions:

$$\frac{\varepsilon_w \varepsilon_0}{r^2} \frac{d}{dr} \left(r^2 \frac{d\phi}{dr} \right) = -e \left[c_+(r) - c_-(r) + c_{H^+}(r) \right] \tag{5}$$

with the ion concentrations given by

$$c_\pm(r) = c \exp\left(-\left[\pm e\phi + U_\pm(r) \right] / kT \right) \tag{6}$$

Here, ϕ is the self-consistent electrostatic potential and $U_\pm$ is the interaction potential experienced by the ions. For the other symbols, see the original papers [9,43,46].

Some of the main results of these works are the following:

1. Two different quantities, bulk pH (minus the logarithm of the chemical potential) and surface pH (minus the logarithm of the electrochemical potential of a hydronium ion at the protein-solution interface), do exist.
2. While chemical potential is constant, the electrochemical potential changes near interfaces [47]. Surface pH is the quantity that influences the number of acid and basic charge groups on a surface. Bulk pH is, in general, quite different compared to surface pH near a protein surface.
3. The addition of small amounts of an anion having a high excess polarizability (SCN^-) produces the same effect (in terms of pH variation) as higher amounts of an anion with a low excess of polarizability (Cl^-).

The difference between the bulk pH and the surface pH (or charge) of a protein depends on the anion present. In particular, highly polarizable anions such as SCN^- are strongly attracted, because of dispersion forces, towards the protein surface. This leads to more hydronium ions near the surface, or more precisely, to a higher surface electrochemical potential, and more bound hydronium ions (higher charge). This attraction can change sign depending on the dielectric properties of the surface with which the ion interacts.

Buffer and protein solutions behave similarly. Thus, for a buffer solution (e.g. phosphoric acid/phosphate), the fraction of neutral and negatively charged species depends on the pK_a values and on the bulk pH. However, it depends also on the background salt solution via the "surface pH" or physico-chemical environment near the buffer anion. Standard textbooks on pH in buffer and salt solutions never take into account any ion specificity (except a number of fitting parameters that should be different for each new situation).

As the salt concentration increases, the concentration of positively charged cations increases in the local region near the negatively charged phosphate ion. This leads by electrostatics to fewer hydronium ions bound to the phosphate ion and more in solution. Thus, the bulk pH decreases with added salt. This effect is even more pronounced at high pH value where a higher fraction of HPO_4^{2-} is present and so more cations are attracted (and hence less hydronium ions). However, the pH decrease will also be different for different ion pairs. If SCN^- and Cl^- ions are compared, the more polarizable SCN^- anion experiences larger attractive ionic dispersion potentials towards the phosphate ion than does Cl^- ion. Bulk pH should be higher in NaSCN solution than in a NaCl solution since there will be more anions near each phosphate ion with SCN^- anions than with Cl^- anions. This result is shown in Fig. 9a only and for salt concentration higher than 1 M. Following this reasoning, one should expect that SO_4^{2-} anions, because of their high charge and low excess polarizability, lead by electrostatics to a more pronounced pH decrease. This is not observed (Fig. 9) as will be remarked on further below. In fact, for phosphate buffer at 200 mM, no significant differences are found as a result of the different salt addition (Fig. 9b).

Even in the presence of a cationic buffer such as Tris-HCl, what is expected is that the effect of increasing the anion excess polarizability is usually to increase the bulk pH. These trends (both the main electrostatic part and the ion specific part) are in general agreement with our measurements with Cl^- and SCN^-. However, it is remarkable that the less polarizable (but with two net charges) SO_4^{2-} anion gives the highest pH increases.

In order to explain results obtained with Na_2SO_4, we should reemphasize that the interpretation of the experimental measurement of pH is an open question. As reported below in Sect. 4, this is real and an unresolved matter that may have still further ramifications.

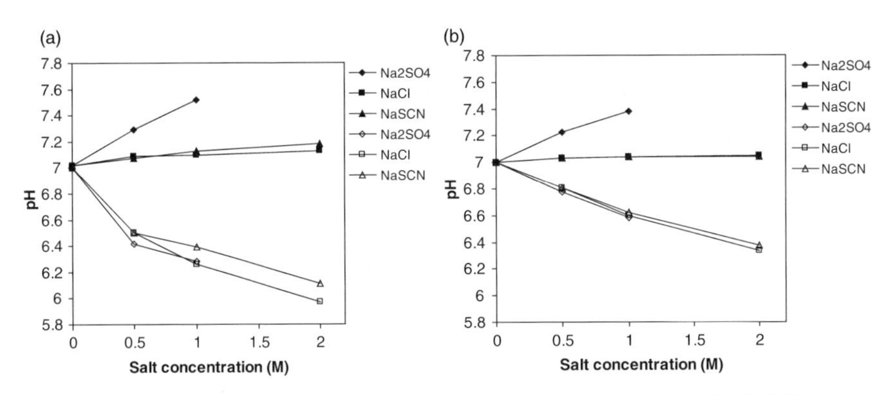

Fig. 9 Measured pH of buffer solutions (**a**): 5 mM; (**b**): 200 mM at initial pH 7 with different concentrations of added salt. Closed symbols refer to Tris-HCl buffer; open symbols refer to phosphate buffer

2.2.4 Structure of the Lipase from *Candida rugosa*

In order to rationalize our results, we recall what is known about the structure and the active site of *Candida rugosa* lipase. With this in mind, we can then see how

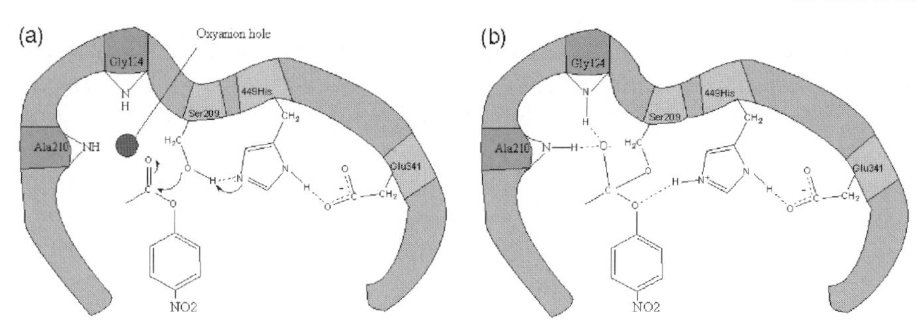

Fig. 10 Schematic representation of the active site of*Candida rugosa* lipase. (**a**) Binding of the substrate (p-nitrophenyl acetate) by the Ser209 with the assistance of Glu341 and His449. (**b**) Formation of the tetrahedral intermediate and stabilization of the oxyanion

competing dispersion or NES forces acting on ions can accommodate the specific effect of different anions on the measured enzymatic activity.

The lipase from *Candida rugosa* (CRL) comprises 534 amino acids with M_r of approximately 60 kDa. CRL is a single-domain molecule and belongs to the family of α/β hydrolase fold proteins. The enzyme shows about a 40% amino acid sequence identity to *Geotrichum candidum* lipase and about 25% equivalent to acetylcholinesterase. Conserved residues include the catalytic triad, disulfide forming cysteines, and some salt bridges. The active site (see Fig. 10) is constituted by the aminoacids Ser209, His449 and Glu341 [48]. This site is buried from the solvent by a single surface loop (flap). The structure of this enzyme has been resolved in two conformations: one where the flap occludes the active site from the solvent (closed form) [49]; the other where the active site is fully available to the solvent (open form) [48].

The loop has an elongated shape and lies flat on the protein surface above the active site. It encompasses a distorted helical turn and a α-helix, and has an amphipatic character. It is hydrophobic on the side facing the protein and interacts with the hydrophobic residues surrounding the active site. On the upper side, exposed to the solvent, it is hydrophilic [49]. Besides the catalytic triad, as will be explained below, the oxyanion hole also has a fundamental role in the catalysis, since it stabilizes the transition state. This is formed by the NH groups of Ala210 and Gly124 [48].

2.2.5 New Insights on Specific Anion Effects on Enzymatic Activity

At the active site of CRL, three amino acid residues are involved in the catalytic mechanism. The Ser209 residue is neutral whereas the other two residues are charged; in particular His449 (pK_a6.04) is cationic and Glu341 (pK_a4.07) anionic. In the first step of the catalytic path, the lone pair of the :OH group from Ser209 attacks the ester bond of the p-NPA to form a tetrahedral intermediate. This electron pair is readily available because of the presence of the Glu341 and the His449 residues (Fig. 10a). A tetrahedral intermediate with an oxyanion is formed; this last is located in the "oxyanion hole". It is stabilized by the formation of two hydrogen bonds with the NH groups of Ala210 and Gly124 (Fig. 10b).

The catalytic activity is strongly affected by surface pH (which determines the ionization state of the amino acids) of the active site. The mechanism is inhibited when both residues Glu341 and His449 are protonated, and cannot then assist the oxygen of Ser209 in its nucleophilic attack on the carbonyl carbon of the ester. This prevents the formation of the tetrahedral intermediate that involves His449 (proton transfer from Ser209) and Glu341 for the delocalization of the charge.

On the basis of the theoretical results obtained by Bostrom et al. [43, 46], the presence of a strongly polarizable SCN^- anion at high concentrations should lead to a higher amount of hydronium ion bound to surface amino acid residues. When the *Candida rugosa* lipase is in its active conformation, the active site is fully available to the solvent [48], so that adsorption of SCN^- is possible and likely to occur. This should lead to the protonation of His449 and Glu341 that can no longer assist the Ser209 in the nucleophilic attack of the carbonyl group of the p-NPA.

In general, protonation of His residues correspond to the already observed [50] apparent pK_a shift that was supposed to be anion specific. Taking into account the effects of dispersion forces on pK_a theoretical calculations of His residues have already been shown to account for the mechanism at the origin of the observed effect [51].

Besides this effect, an additional phenomenon might be responsible for the dramatic decrease in enzymatic activity observed with SCN^- anions. The native conformation of an enzyme is determined by a complex interplay of many body molecular forces subsumed under names like ionic, hydrogen bond, many body dipolar and dipole-induced dipole and van der Waals interactions. These forces provide stability to the enzyme under physiological conditions and prevent deleterious conformational changes that could cause deactivation. It is likely that SCN^- adsorption is able to interfere with the hydration of groups involved in these interactions at the enzyme surface level even at the active site. This could cause a distortion of the active site i.e. of the amino acids of the catalytic triad and those forming the oxyanion hole. This distortion should modify the cavity and so render the stabilization of the transition state impossible. This effect, if confirmed, should lead to the dramatic diminution of enzymatic activity as shown in Figs. 7–8.

The destruction of enzyme structure caused by chaotropic anions is an explanation commonly given to justify enzyme inactivation [11]. What is new here is that the direct effect of anions can be traced to the anionic adsorption driven by the many body dispersion forces rather than an effect mediated by changes in bulk water structure [39–41].

On the other side of the Hofmeister series, the kosmotropic anion SO_4^{2-} implies that it interacts mainly via electrostatics alone. The activating-deactivating phenomena depend on the solution buffer. In this case, the effect on the activity caused by SO_4^{2-} addition is likely to be related, at least at a first approximation, to the bulk pH shift. This is true, for example, for phosphate buffer 200 mM where the addition of 1 M Na_2SO_4 gives a pH shift from 7.00 to 6.62 (see Fig. 9b) and an enzymatic activity of 289 $\mu mol\,min^{-1}\,g^{-1}$ (Fig. 7b). This pH shift goes in the direction of the maximum in the activity versus pH curve (pH = 6.52; enzymatic activity = 268 μmol min^{-1} g^{-1}) in Fig. 6b.

In the case of 200 mM Tris-HCl, the addition of 1M Na_2SO_4 gives a pH shift from 7.00 to 7.38 (Fig. 9b) and an enzymatic activity of 188 μmol min^{-1} g^{-1} (Fig. 8b). If this pH shift in the activity versus pH curve is considered (Fig. 6b), an enzymatic activity = 140 μmol min^{-1} g^{-1} at pH = 7.44 is obtained. The different effect of sulfate anion (activation/deactivation) observed with the different buffers confirms the importance of the choice of weak electrolytes used to set the bulk pH. The choice of buffer is supposed not to be influential, but clearly it is.

If these explanations are correct, the effect induced by the HS neutral anion Cl^- is partially due to a combination of nonspecific electrostatic forces and ion-specific NES forces. This is also tuned by the solution buffer, with competition for the active site between chloride and buffer ions being explicit. That this is probably so receives support from the observation that the effect of high buffer concentration (200 mM) is also explainable in terms of Hofmeister series. Indeed, when buffer concentration is low (5 mM), its weak buffering capacity does not allow to reach a single maximum. It is however remarkable that the three buffers produce similar maximal activities although at different pH values (cfr. Fig. 6a). When buffer concentration increases, the higher buffering capacity allows to reach the real value of optimal pH (around 6.5). At this point, a Hofmeister effect appears giving different activities, namely a high value for the kosmotropic phosphate and citrate anionic buffers and a low value for the chaotropic Tris-HCl cationic buffer (cfr. Fig. 6b).

2.3 *Inferences from Hofmeister Effects Observed in Enzyme Activity*

The present works demonstrate explicitly the specific effects of weak and strong electrolytes on enzymatic activity of a lipase. Previously, the consensus has been that the former acts only as buffers, i.e. to set pH only, and that the latter acts as ionic strength regulators, i.e. to modify electrostatic forces of association only. Here, we have shown that this picture is much too simple. Both buffers and salt anions have specific effects that fall within the ambit of the Hofmeister series phenomena. This is not a trivial phenomenon. It is one that conventional biochemistry text books have neglected even though Hofmeister effects have been known for more than 100 years and first demonstrated with protein solubilities. The reason is evidently due to the circumstance that the underlying theories of physical chemistry have not been fully satisfactory and predictive. They have been mainly limiting laws valid only for dilute solutions and inadequate for the job.

Recently, some of the flaws in classical theories have been exposed [52]. Incorporation of many body NES forces into theories is providing new insights into Hofmeister phenomena. These insights can be used to rationalize several phenomena both in biochemistry and in colloidal chemistry problems [53, 54].

3 Colloid Stability: Limits of the DLVO Theory

The stability of colloidal systems is determined by a delicate interplay of various intermolecular forces. Supramolecular microstructures, long-range ordered liquid crystals and also some simple biological systems are often used as models that reflect the role of different intermolecular forces in stabilization processes and in the evolution of microstructure. For more than half a century, the DLVO theory [55,56] has underpinned the intuition of colloid scientists on these forces. The apparent successes of DLVO theory derive from its simplicity that allows the extraction of the essential physics of the problem of lyophobic colloid stability. Its starting point is the ansatz that colloidal particle interactions (the stability or coagulation of dispersions) is determined by the balance of two separate forces: the double layer repulsion and the van der Waals attraction.

But the two forces are intimately coupled. The separation of forces is an invalid approximation [52]. This limits the applicability of the theory to low electrolyte concentrations ($1 \times 10^{-3} - 5 \times 10^{-2}$ M). Fig. 11 summarizes the two relevant equations describing the repulsive and the attractive potentials between two particles.

DLVO theory loses its predictive ability in a huge number of colloidal phenomena that are dominated by the problem of ion specificity, or Hofmeister effects. This problem, extant for more than a century is one for which, indeed, the original DLVO theory did not claim any competence. Moreover, the further approximations are that: (1) the solvent is a bulk continuum, unperturbed by an electrolyte; (2) that no profile

Current theories: DLVO

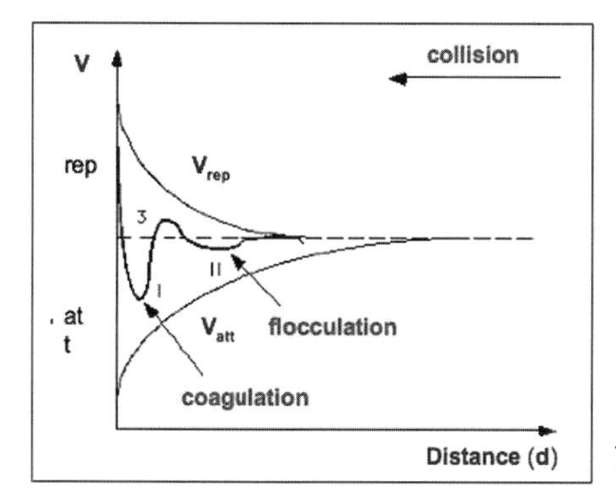

Electrostatic repulsive Potential

$$V_{rep} = 2\varepsilon\varepsilon_\circ \Psi_\circ^2 \kappa \exp(-\kappa d)$$

$$\kappa \propto (2I)^{0.5} \quad I = \text{ionic strength}$$

Attractive Potential

$$V_{att} = -\frac{H\, r_1 r_2}{6d(r_1 + r_2)} \quad (\textit{between two spheres})$$

H = Hamaker constant

(*function of sphere density*)

Fig. 11 DLVO may explain colloidal stability in very diluted regime, but not HS effects or ion-specific induced superactivity in enzymes

of surface induced solvent order (hydration) exists at an idealized interface; and especially (3) that no perturbation of this hydration occurs due to interaction with adsorbed hydrated ions. (The first and third assumptions remain open. The second is justified thermodynamically by the theorem that for the evaluation of free energies, it is sufficient that the distribution function, i.e., profile of order is correct only to zeroth approximation).

Theoretical developments have treated the electrostatic double layer interactions by a nonlinear theory of electrolytes. The second kind of (nonelectrostatic, NES) forces have been handled by the linear theory of Lifshitz that in principle deals with all many-body quantum mechanical fluctuation forces. This procedure, with the one force treated by a non linear theory, and the others by a linear theory violates both the Gibbs adsorption equation and the gauge condition on the electromagnetic field [1,52]. As a consequence, it can be shown that specific ion effects are missed. The words "specific ion effects" then are a mnemonic that disguises the fact that there has been no encompassing theory of systems involving aqueous electrolytes except in the limit of extreme dilution (the situation has been reviewed in a recent special issue of Current Opinion Colloid Interface Science [5]). The problem is endemic and occurs not just for colloidal interactions, but also for the Born energies of transfer, correlation free energies of electrolytes (Debye Huckel theory), and interfacial tensions. When the defects are remedied by treating both kinds of forces at the same level in a consistent nonlinear theory, specific ion effects do show up and apparently predictively [37]. There are many more recent theoretical studies. Indeed co- and counterion specificity manifests itself in a variety of processes too numerous to list. They range all the way from bacterial growth [57], enzymatic activity [14] as shown in Sect. 2, to self-assembly of surfactants [58, 59], to hydrophobic chromatography [33] and the effects can usually be ordered (at least at a qualitative level) within the well known Hofmeister series [32]. The series is often reversed and depends on surface or substrate, and buffer type, a fact that alone implicates the NES forces.

While progress has been made by working just at the level of continuum solvent approximation, the major source of the effects is still debated. And reasonably so. For, the extension of theory just discussed does not include bulk salt and surface induced water structure. Nor does it include effects of dissolved gas [60, 61].

To account for ion specificity, two parallel lines of thought have been followed. The two approaches apparently sit in apposition and opposition to each other. One assigns to ionic hydration, the entire responsibility and carriage of the ion specific effects [62, 63]. The other insists on the fact that ions near an interface must experience a potential due to (NES) dispersion forces in concert with the classical electrostatic potential [37, 52]. Basically, the first approach ascribes ion specificity to ionic influence on water structure, the other to relative ionic affinity to interfaces. Both effects are certainly operating. The question is which dominates. The first approach is more complicated as hydrated ion-ion interactions or hydrated ion-surface interactions necessarily involve more, or less, overlap of hydration shells that leads to the concept of "hard" and "soft" ions, which itself goes back to the "civilized" model of electrolytes of Stokes [2]. Although the second focuses on surface-ion interactions via forces not included in conventional theory, the NES ionic and solvent frequency-dependent dielectric susceptibilities (as exemplified by excess polarizabilities) play a role in both theories. The (self) interaction of a bare ion with water in the presence of its neighbors and/or an interface via NES interactions [64–66] gives rise to strong or weak hydration and the characterisation of ions as kosmotropic or chaotropic. These hydrated or "dressed" ions then experience further specific NES dispersion potentials with an interface directly, and with its profile of hydration. The two apparently different approaches are consistent, the second, in fact, including at least a part of the first. Which effect dominates in a particular case is a matter of dispute [9]. This fact renders somewhat ambiguous the attribution of experimental results to one approach (bulk effects), as opposed to the other (surface effects). An emblematic example is the following: Dialkyldimethylammonium bromide (DDAB)/water/hydrocarbon mixtures show a large reverse (water-in-oil) microemulsion region [67–69] whereas changing the DDAB counterion with sulfate (DDAS), a normal (oil-in-water) microemulsion is found [58, 59]. Titration of the DDAS/water/hydrocarbon microemulsion with a NaBr solution induces a transition back to the reverse phase system. Surprisingly, the change in curvature occurs at a mole fraction of bromide equal to 0.4 [59]. Typical counterion (sulfate) concentrations are around 1 M. Such a result clearly indicates a marked preference of bromide to the interface with respect to sulfate. From an electrostatic point of view, this is, of course, plainly absurd. The divalent sulfate ion ought to be the clear winner in a competition for the charged surface with the univalent bromide ion. This behavior can be attributed to the stronger attractive NES interaction between bromide and the interface as determined by its higher polarizability. Although dispersion forces must be called into question, there are no evident reasons (experimental results) to argue against the fact that bromide preference for the charged surface could also be due to its chaotropic behavior (sulfate is a kosmotropic ion). Therefore, in such an example, both proposed theories hold. (Earlier preliminary studies of coion and counterion effects in this system showed up some extraordinary changes in the phase diagrams [70]).

At the present time, there is progress in extending solution theory further to include hydration effects and hydration shell overlap within the same extended nonlinearized continuum Lifshitz–double layer formalism. Once that is done, any apparent conflict between apparently opposing schools of thought disappear.

4 Salt Effects on pH of Buffered Solutions

The presence of a background electrolyte, its type and concentration, can affect the pH of a solution [71–76]. With a glass electrode, the measured pH changes with added electrolyte, buffered or not. The changes depend on salt type [76]. Such effects are not accommodated within the standard IUPAC Recommendations on pH measurements and procedures [77]. These recommendations rely on the extended Debye-Hückel theory of electrolytes and the Nernst equation (see Sects. 5.1 and

IONIC STRENGTH
Adding an inert salt increases the solubility of an ionic compound.

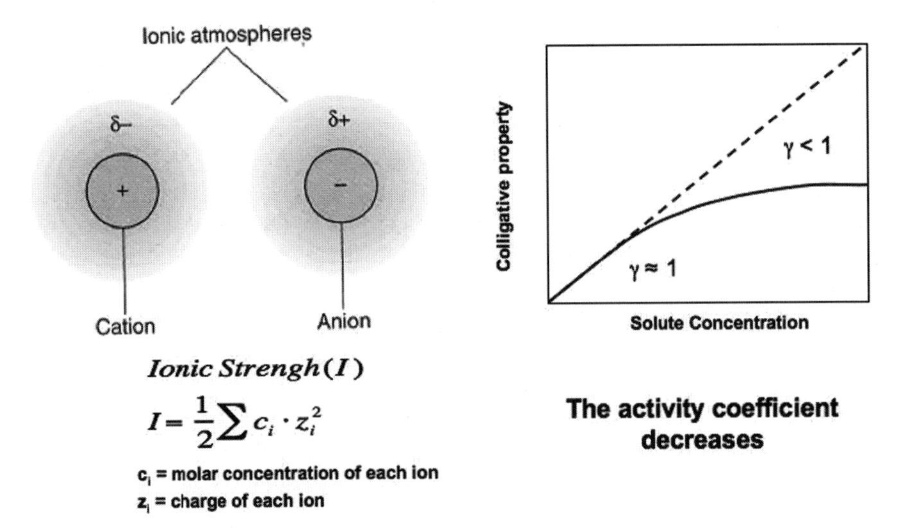

Fig. 12 Activity coefficients and ionic strength: the net attraction between the cation and its ionic atmosphere and between the anion and its ionic atmosphere is smaller than it would be between pure cation and anion. As a consequence, the solubility increases

11.1 in that reference). Neither theoretical assumptions embrace specific ion effects. For the best up-to-date account of the situation on hydronium activity determination, see [78].

Before describing the effects of added salts to buffered solutions, it is worth recalling the basic concepts of ionic strength: this is shown in Fig. 12. The equation reported in Fig. 12 is valid only in very dilute regimes. Indeed, it is known that adding an inert (neutral) salt increases the solubility of an ionic compound since the net attraction between the cation and its ionic atmosphere and between the anion and its ionic atmosphere is generally smaller than it would be between pure cation and anion. In other words, the activity coefficients decrease; therefore, important corrections are to be used for real concentrations. Fig. 13 summarizes the main equations used to calculate the activity coefficients. It should be noted that only the Debye–Huckel extended theory introduces a specific ion parameter, that must be determined experimentally via a fitting procedure.

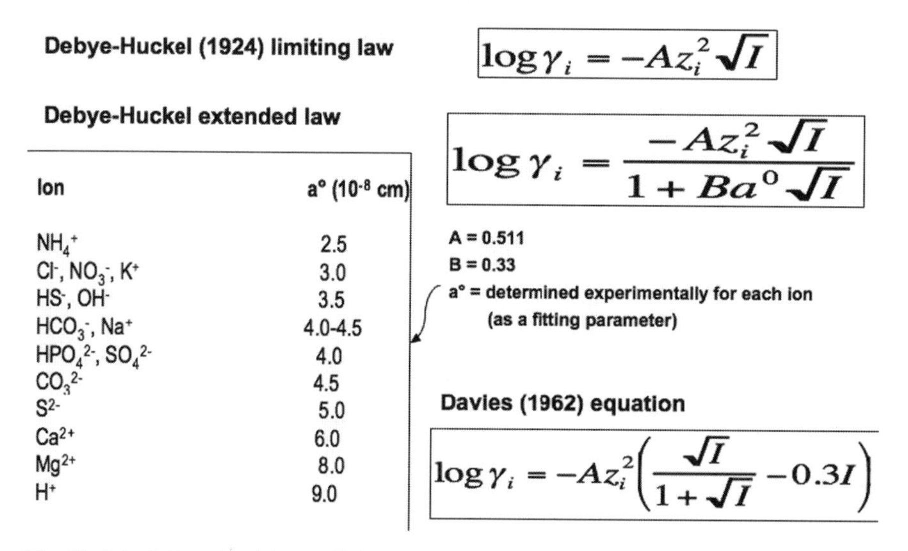

Fig. 13 Calculation of activity coefficients

4.1 Phosphate and Cacodylate Buffers (Sodium Salts)

Turning the attention to pH measurements of buffered solutions in the presence of added electrolytes, it should be mentioned that the absence of the buffer prevents reliable determinations of pH values. This is due to CO_2 dissolution whose extent

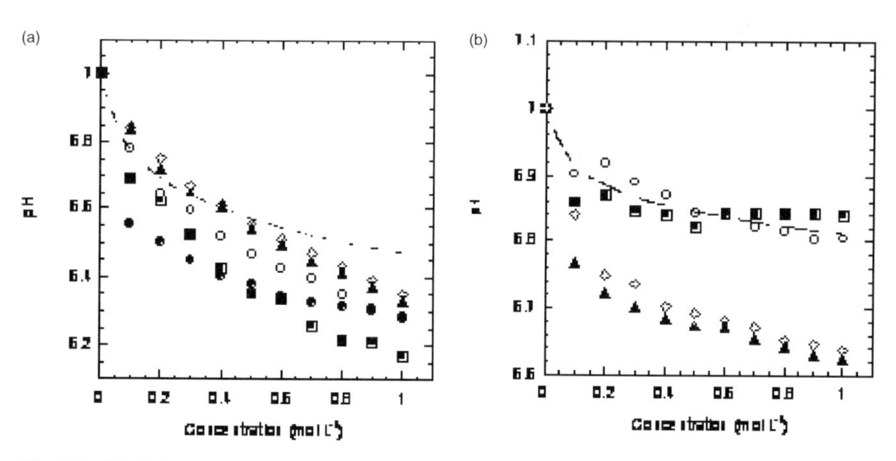

Fig. 14 pH of phosphate (**a**) and cacodylate (**b**) buffers (5 mM, initial pH 7) with increasing concentration of salts: NaCl (◊); NaBr (▲); $NaNO_3$ (○); NaI (•); $NaClO_4$ (□); calculated pH from DH limiting law (dashed line). $SD_{max} = \pm 0.02$ pH units

is determined by the added salt concentration but cannot be precisely quantified. Nevertheless, reverse HS effects clearly emerge from experimental data.

Hence, sodium phosphate and sodium cacodylate buffers 5 and 20 mM, at pH = 7, were used as starting solutions. To these, the various sodium salts were added. pH values were measured at different salt concentrations. Figs. 14a and 14b show the results of pH measurements as a function of sodium salt concentration for starting solutions of 5 mM phosphate and cacodylate buffers at initial pH = 7.

The dashed lines indicate the pH values calculated according to the extended Debye-Hückel (DH) theory (see Eqs. in Figs. 12–13 and section below).

The first remark is that in the presence of a buffer at pH 7, the addition of salts causes a significant decrease of the measured pH values as concentration increases. Different pH changes are recorded in the two buffers for the various sodium salts. The efficacy of anions in changing pH follows different sequences. The effects are strongly dependent on the nature of the buffer.

At high salt concentration (1 M), ΔpH values in the range 0.60–0.85 and of 0.2–0.4 pH units are observed for phosphate and cacodylate, respectively. Phosphate buffers induce a larger decrease of pH values than cacodylate. The extended DH theory calculations provide a trend qualitatively similar to those experimentally observed (see the discussion below). For the 5 mM buffers, the anion effectiveness in changing pH is: $ClO_4^- > NO_3^- \approx I^- > Br^- > Cl^-$ for phosphate buffer, and $Br^- > Cl^- > NO_3^- > ClO_4^-$ for cacodylate buffer, with an inversion in the series.

It is worth recalling that – in terms of the kosmotropic/chaotropic model – the phosphate anion itself is probably the most efficient kosmotrope, whereas the added anion having the highest chaotropic effect (ClO_4^-) is the most effective in changing the measured pH. Thus, the anions show a decreasing effectiveness from chaotrope to kosmotrope in the Hofmeister anion sequence. For the phosphate buffer, a clear crossover between NaI and $NaClO_4$ occurs around 0.4 M salt concentration (Fig. 14a). The same effect is observed between $NaNO_3$ and $NaClO_4$ in cacodylate buffer (Fig. 14b). This effect does not appear in the 20 mM phosphate buffer (see Fig. 16). The effect of buffer concentration was further investigated for NaCl and $NaClO_4$ salts (see par. 4.3).

4.2 *Reversal of Effects with Potassium Salts*

The measurements shown in Fig. 14a were repeated replacing sodium with potassium. Because of its low solubility, potassium perchlorate was replaced by potassium thiocyanate.

The pH measurements were performed up to 2 M concentration of salts. A much smaller decrease of pH was observed with potassium than with sodium phosphate buffer. And, considering the data measured at 1M concentration, the effects are in the order $Br^- > Cl^- = NO_3^- > SCN^- > I^-$ as shown in Fig. 15. That is, replacing Na^+ with K^+ brings about a reverse Hofmeister series effect. It seems difficult to assign this finding to a bulk effect, and presumably a mechanism involving specific surface adsorption of cations is indicated [9].

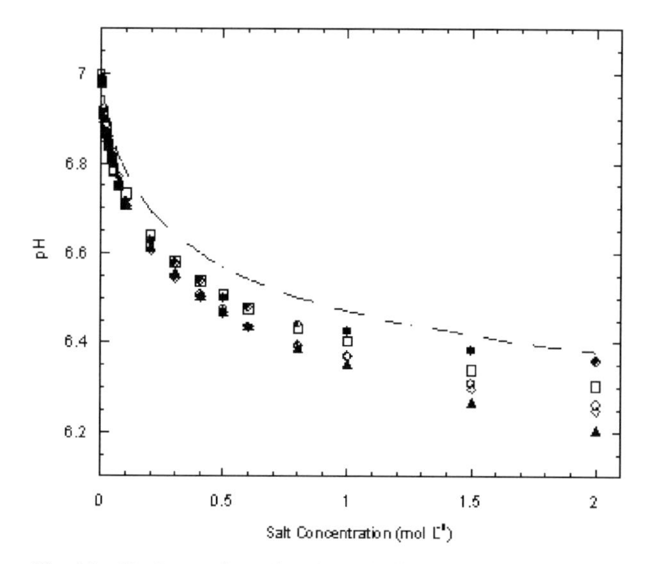

Fig. 15 pH of potassium phosphate (buffer 5 mM, pH 7) with increasing concentration of salts: KCl (◊); KBr (▲); KI (•); KNO_3 (○); KSCN (□); calculated pH from DH limiting law (dashed line). $SD_{max} = \pm 0.02$ pH units

4.3 Role of Buffer Concentration

Phosphate buffer solutions 0.1, 1, 5, and 20 mM were used. Results are shown in Fig. 16a for NaCl and Fig. 16b for $NaClO_4$.

Besides the expected major effect of $NaClO_4$ salt, the role of buffer concentration can be clearly seen. At 0.1 mM, the buffer shows a fluctuating, but markedly decreasing trend for both salts. ΔpH values of about 0.9 and 1.3 are measured for Cl^- and $ClO_4{}^-$ anions, respectively. As expected, the ΔpH effects decrease significantly with increasing buffer concentration.

The more dilute the buffer solution, the lower the recorded pH value, regardless of the electrolyte concentration.

Finally, we remark that cations also seem to play a key role as demonstrated by the different behavior observed in Na^+ and K^+ phosphate/salt systems.

4.4 pH Values and Intermolecular Forces

Besides any speculation that might be made on the interpretation of these results, it is clear that a delicate balance related to the different composition of buffers and salts, and to their respective concentrations, plays a subtle role in determining the measurement of pH of the solution. All these variables affect pH values significantly. At least what we measure. This seems not to have been widely recognized. Whether the measurements reflect a real or only an apparent change in pH seems to be an important open question. It is perhaps not too surprising that in the classical theory that underlies interpretation of the pH measurement, NES forces acting on ions are missing.

The observations have relevance to biological problems. This is so for two reasons. The changes in pH measured – whether real or apparent – depend on salt concentration, i.e. ionic strength or Debye length. In a real biological situation, for example physiological saline roughly 0.15 M, it might be expected that these effects are small. But, in fact, a real system – the cytoplasm – contains a significant amount of proteins, carbohydrates, and other multivalent species. For such a system, even a very small amount of multivalent ions increases the Debye length dramatically [79–81].

Not only the measured pH is influenced, dramatic changes in the enzymatic activity were found as a result of added electrolytes in different buffers [14,19,57,82–84].

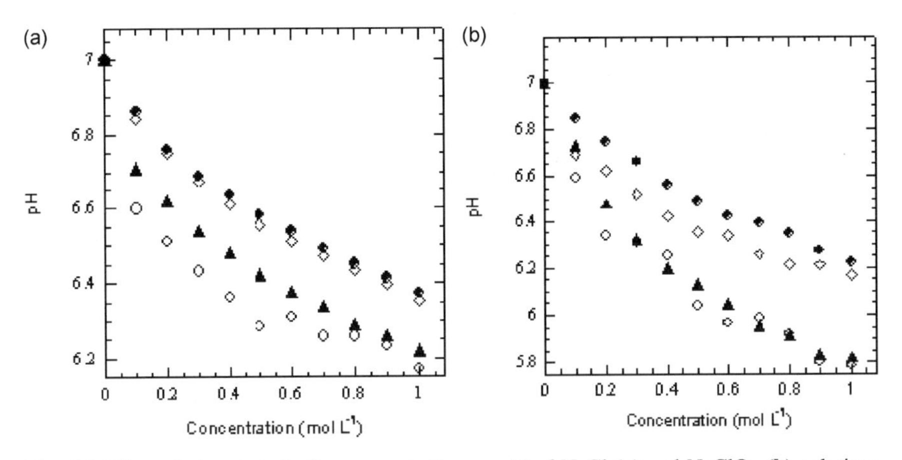

Fig. 16 Effect of phosphate buffer concentration on pH of NaCl (**a**) and $NaClO_4$ (**b**) solutions. 0.1 mM (○); 1 mM (▲); 5 mM (◇); 20 mM (•). $SD_{max} = \pm 0.02$ pH units for buffer concentration 5 and 20 mM and $SD_{max} = \pm 0.05$ pH units for buffer concentration 0.1 and 1 mM

4.5 *Correlation Between pH and Other Physico-chemical Properties: Hofmeister Fingerprints*

We summarize our results in terms of direct or reverse Hofmeister series (HS). The usual, direct Hofmeister series is:

$$H_2PO_4^- > SO_4^{2-} > F^- > Cl^- > Br^- > NO_3^- > I^- > ClO_4^- > SCN^-$$

Then, in terms of effectiveness in modifying pH, we observe

pure water; Na^+ salts	$I^- > ClO_4^- > Br^- > Cl^- > NO_3^-$	reverse HS
Na^+ phosphate 5 mM; Na^+ salts	$ClO_4^- > NO_3^- \approx I^- > Br^- > Cl^-$	reverse HS
Na^+ cacodylate 5 mM; Na^+ salts	$Br^- > Cl^- > NO_3^- > ClO_4^-$	direct HS
K^+ phosphate 5 mM; K^+ salts	$Br^- > Cl^- > NO_3^- > SCN^- > I^-$	direct HS

Phenomena where significant specific ion effects occur usually show correlations when the experimental data are compared to some physico-chemical parameters that are characteristic fingerprints of Hofmeister effects. Typically, some of these quantities are: ion excess polarizability, partial molar volume, molar refractivity, surface tension molar increment, Gibbs free energy and entropy change of hydration, lyotropic number, Jones-Dole viscosity B-coefficients, Setschenow constants, entropy change of water, and so forth. Since they depend on the chemical nature, these quantities provide a marker for each ionic species or ion pair [1].

Regular trends are found when the experimental pH values are related, at fixed concentrations, to the partial molar volume (ν_s) of the anions, the molar refractivity (R_s) and the surface tension molar increment (σ). For a more detailed presentation of these parameters, see [1, 27]. Table 1 summarizes those parameters along with the pH values measured in the presence of the various salts, at 0.5 and 1 M concentration, and in 5 mM and 20 mM phosphate buffers.

Fig. 17 shows the change in the measured pH value as a function of the partial molar volume (ν_s) of each anion. Remarkably, an almost linear trend is observed. The pH decreases with increasing ν_s, and R_s, and with decreasing σ. In terms of the

Table 1 pH values of phosphate buffer solutions in the presence of different electrolytes (0.5 and 1M), partial molar volume (ν_s, cm^3/mol), molar refractivity [85] (R_s for 1M salt solutions, cm^3/mol), and molar surface tension increment (σ, $mN \cdot L/m \cdot mol$) values for each anion

pH					ν_s	R_s	σ
	[salt]=0.5 M		[salt]=1 M				
	[buffer]=20 mM	[buffer]=5 mM	[buffer]=20 mM	[buffer]=5 mM			
H_2O	7.0	7.0	7.0	7.0			
Cl^-	6.6	6.6	6.4	6.3_5	16.6	8.51	1.63–2.21
Br^-	6.5	6.5	6.3	6.3	23.5	12.13	1.31–1.83
NO_3^-	6.5	6.5	6.3	6.3	27.7	11.02	0.89–2.08
I^-	6.5	6.4	6.3	6.3	35.0	13.72	1.08–1.23
ClO_4^-	6.5	6.3_5	6.2	6.2	43.0	19.25	0.22–0.62

kosmotropic/chaotropic language, the pH is less affected by the harder, less polarizable kosmotropes (Cl^-). It consistently decreases with the softer, more polarizable chaotropes (ClO_4^- and I^-). These properties (ν_S, R_S, and σ) are directly related via (mainly) polarizability of anions. This is so for hydration or self free energies of ions [64]. It is also so for changes in Born and dispersion self energies due to different ion-substrate interactions, that are given by the same theoretical formalism of Lifshitz [52, 64–66].

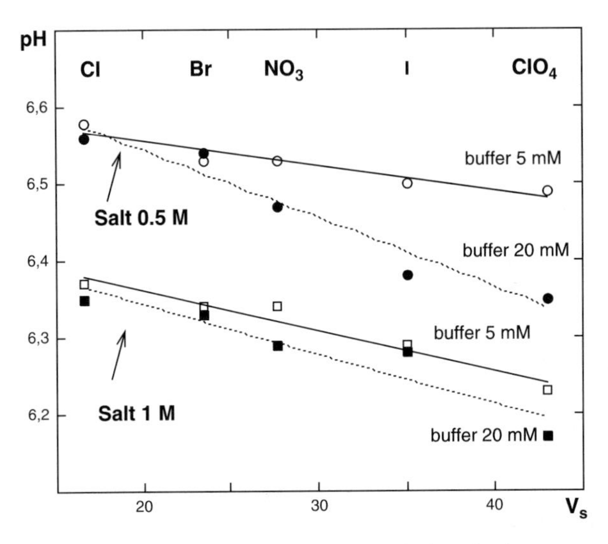

Fig. 17 pH values of phosphate buffer solutions in the presence of different electrolytes, as a function of the partial molar volume. •: [buffer] = 20 mM and [salt] = 0.5 M; ○: [buffer] = 5 mM and [salt] = 0.5 M; △: [buffer] = 20 mM and [salt] = 1 M; ▲: [buffer] = 5 mM and [salt] = 1 M. Partial molar volumes from [85]

We can also infer that the same highly specific electrodynamic NES interactions must participate in interpretation of the salt- and buffer-dependent pH change.

These observations, together with the reversals with buffer type and cation, suggest that the phenomenon is due to a delicate interplay of different intermolecular interactions. These interactions are related to polarizabilities and ionization potentials, and to corresponding dielectric properties of substrate and water.

We now discuss how NES forces might affect pH via bulk and surface interactions between ions in the solution and between ions and glass electrode surface respectively.

4.6 Bulk Interactions

A direct bulk effect of the activities ($a = \gamma \times c$) of the acid/base pairs ($H_2PO_4^-/HPO_4^{2-}$ and $(CH_3)_2AsOOH/(CH_3)_2AsOO^-$) must be considered. The standard analysis used to calculate the activity coefficients of the buffers in the presence of salts is here resumed by giving the most important formulae.

From the equilibrium:

$$H_2PO_4^- + H_2O \rightleftarrows HPO_4^{2-} + H_3O^+$$

We can write the expression of the thermodynamic equilibrium constant K_{a2}^0:

$$K_{a2}^0 = \frac{a\left(HPO_4^{2-}\right) \times a(H_3O^+)}{a(H_2PO_4^-) \times a(H_2O)} \tag{7}$$

$$pH = pK_{a2}^0 + \log\frac{\left[HPO_4^{2-}\right]}{\left[H_2PO_4^-\right]} + \log\frac{\gamma(HPO_4^{2-})}{\gamma(H_2PO_4^-)} - \log a(H_2O) \tag{8}$$

Where $\gamma(X^-)$ and $[X^-]$ are the activity coefficient and the molar concentration of the generic anion X^-, respectively. A similar equation can be written for the cacodylate buffer.

From these equations, the expected pH can be calculated, assuming that K_{a2}^0 and the molar concentrations of the acid and its conjugated base are constant. With this assumption, pH changes originate from the variation in activity coefficients with ionic strength I. Activity coefficients are calculated from the extended Debye–Hückel (DH) equation:

$$-\log \gamma_z = Az^2 \frac{\sqrt{I}}{1 + Ba\sqrt{I}} \tag{9}$$

where A and B are constants that depend on temperature and the dielectric constant of the solvent ε; z is the ion charge; and a is an ion size parameter (unknown, subject to best fitting demands); I is the ionic strength.

The dashed lines reported in the Figs. 18 and 19 were obtained using Equations 8–9 (8–9) and assuming $a(H_2O) = 1$. Such equations do not take into account any real ion specificity. More importantly, they do not account for the series inversion observed for phosphate and cacodylate buffers (Fig. 14). Even more problems arise when we replace sodium with potassium (compare data in Figs. 14a and 15). To account for the observed deviations from a purely electrostatic approach, different attempts have been done adding to Equation(9) a first order term (bI), but b is again a fitting parameter [2, 86]. Another approach can be the introduction, in the pH calculations, of the activity coefficients of water obtained from the osmotic coefficients according to the usual equation for an electrolyte solution [2]:

$$\ln a(H_2O) = \frac{vmW_{H2O}}{1000}\phi \tag{10}$$

where v is the number of ions formed from one mole of electrolyte; m is the molality (mol Kg^{-1}), W_{H2O} is the molecular weight of water, ϕ is the osmotic coefficient of the electrolyte. Then, the pH values for NaCl, NaBr, and $NaNO_3$ salts were calculated using also Equation 10 and according to Equation 8, (e.g. the extended DH equation with $pK^0_{a2\,\text{phosphate}} = 7.2055$;. $A = 0.51$ and $B = 0.33$ at 25°C in water; $a =$ 4 Å for $H_2PO_4^-$ and HPO_4^{2-}) [87]. Comparing these new figures with those calculated by the non-ion-specific extended DH theory, the pH variations for the different salts appear in the 3rd or 4th digit in pH units. Clearly, this is irrelevant either for a more reliable prediction of the experimental data or to justify the observed HS effects. Hence, the theoretical curves reported in Figs. 18 and 19 are simply based on the extended DH law. These results suggest that some other considerations have to be involved.

According to Equation 8, the effect of salts on pH can be related to changes in the values of the activity of HPO_4^{2-}, $H_2PO_4^-$, and water. Since a detailed calculation of $a(H_2O)$ from osmotic coefficients shows that its value changes so slightly from one salt to another, its contribution cannot be responsible for the deviations detected. This evidence prompts us to ascribe the main role to the other terms, namely pK^0_{a2} (K^0_{a2}, the dissociation constant of $H_2PO_4^-$) and $\log[a(HPO_4^{2-})/a(H_2PO_4^-)]$. It may be proposed that NES dispersion interactions between the salt anions and the ionic species that participate in the acid/base equilibrium, can lead to a delicate but non-zero variation of the ratio $K^0_{a2} \times a(H_2PO_4^-)/a(HPO_4^{2-})$. The balance of this

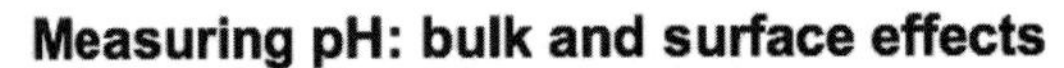

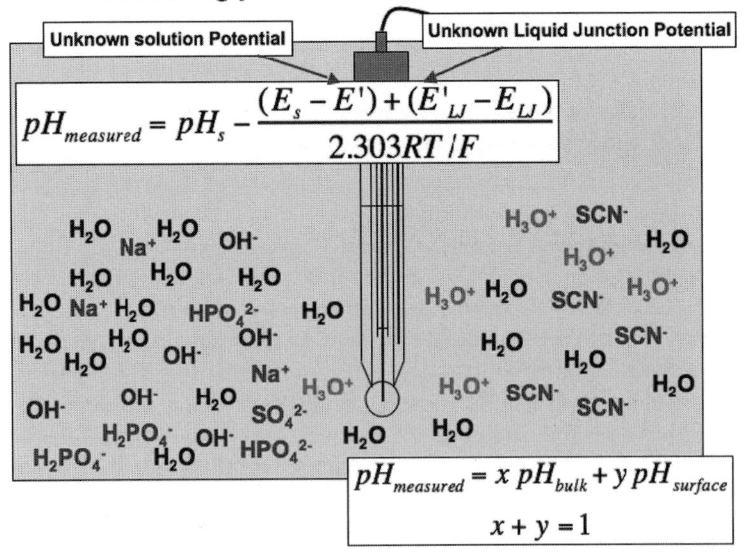

Fig. 18 Phenomena at the glass electrode surface. Ion specific effects can arise from ion specific adsorption at the solid–liquid interface, and from different diffusion velocity across the membrane.(ES and NES forces)

To account for ion specificity
ES + NES forces

In the presence of an electrolyte M^+X^-
The equilibrium HA <=> H^+ + A^- and then the pH
are altered by an interplay of attractive and repulsive interactions
that occur either in the bulk or at the solid-liquid surface (SiO^-)

	Bulk		Surface	
INTERACTION	ES	NES	ES	NES
attractive	H^+/A^-	X^-/A^-	SiO^-/H^+	SiO^-/X^-
attractive	H^+/X^-	X^-/HA	SiO^-/Me^+	HA/SiO^-
attractive	Me^+/A^-			A^-/SiO^-
repulsive	H^+/Me^+		A^-/SiO^-	
repulsive	X^-/A^-		H^+/Me^+	

NES forces are higher for soft ions and electron rich neutral molecules
NES forces are lower for hard ions with one or more charges

Fig. 19 ES and NES forces along with bulk and surface interactions during pH measurements

variation and the change in $a(H_2O)$ would determine the final pH. This hypothesis relies on the fact that salts can change the ionization state of a weak acid [88–90]. What is new here is that NES dispersion forces can be at the origin of such phenomenon. In this context, for instance, it is explained why chloride, that can establish hydrogen bond interactions with the monovalent biprotic species $H_2PO_4^-$, produces the lowest change in pH. Instead, a species less active in producing hydrogen bonds but with a larger polarizability, such as I^-, will stabilize the monoprotic divalent HPO_4^{2-} ion, with a consequent larger effect on pH.

4.6.1 Problems with Current Explanations of Bulk Phenomena

Let us focus now on the results obtained in 5 mM buffers. Figs. 14a and 14b show two main results: (i) The measured pH is not constant but decreases with increasing salt concentration; (ii) The pH decrease is ion specific and follows the Hofmeister series. Tentatively, these results may be accommodated in terms of bulk phenomena. If we consider the range of ionic strength up to 0.1 M, we can, in first approximation, neglect the terms containing the activity coefficients in Equation 8. Even in this case, the measured pH is lower than the expected value of 7. This can be explained in terms of ES forces acting between Na^+, H^+ and their counterparts $H_2PO_4^-$, HPO_4^{2-}. Close to neutrality, $[H^+] = 10^{-7}$ mol/L and $[Na^+] = 10^{-1}$ mol/L, so Na^+ is 10^6 times more concentrated than H^+. Thus, Na^+ will compete with H^+ for interacting with the divalent anion HPO_4^{2-}. This will lead to less H^+ bound to phosphate buffer and more H^+ free in the solution and thus to a lower pH. This phenomenon will increase by increasing salt concentration from 0.1 M to 1.0 M. A similar explanation appears to be responsible for the decrease in pH with salt concentration in the case of cacodylate buffer. The higher pH measured in this case, with respect to the phosphate buffer, should be produced by the weaker interactions between Na^+ and $(CH_3)_2AsOO^-$ as compared to that between Na^+ and $HPO_4{}^{2-}$.

Ion specificity begins to show up strongly at $I > 0.1$ M and increases with concentration. This range of concentration is that where the NES (dispersion) forces begin to dominate [52]. The extended DH theory contains additional ion, ion pair and buffer size parameters in the expression for γ_z (see E(9)). Since no theoretical expression for activity coefficients that considers ion specificity or mixed electrolytes has yet been derived, the availability of such parameters, specific to each case is tantamount to the tautological restatement that pH is buffer and salt concentration dependent. In qualitative terms, dispersion potentials between ions will be different for differently polarizable anions. The effect will give different pH/salt curves for different anions.

4.7 Surface Interactions

Specific ion adsorption at a metal electrode surface is a well-known phenomenon. It was treated in terms of a Langmuir isotherm by Stern in 1924 [91], and the ideas were successively developed by Parsons [92]. It was believed that the adsorption occurs via electronic transfer from the anion to the orbitals of the metallic electrode [93–95]. More recently, a new approach correlating electrochemistry with colloid science emerged. In particular, Conway found a Hofmeister series in the zero charge potential of Hg electrodes. He proposed a new explanation in terms of solvation factor [96]. A succeeding work by Conway [97] correlated the adsorption with the effective ionic radii and hydration volumes. The main conclusion was that electronic effects alone are insufficient to describe ion specificity. Hydration effects in anion adsorption were invoked as the substantial part of the observed specificity of ion adsorption [97].

A scheme to illustrate surface interactions is shown in Fig. 1.

Ion adsorption specificity was also observed at air-water interface where no such electronic effects can be invoked [64, 98, 99]. Clearly, ion adsorption may occur at the glass-water interface as well.

In the past, bulk effects had been expected to account for the general decrease of pH and ion specificity of pH/salt curves. If bulk effects were the only phenomenon responsible for the observed changes, we should expect the same Hofmeister series trends with different buffers for enzyme action [14], and for reverse and direct sequences with proteins [9]. Fig. 14 shows that the Hofmeister series is inverted if cacodylate buffer instead of phosphate is used. In addition, Figs. 14a and 15 indicate a reverse Hofmeister series when sodium is replaced by potassium. These effects may be explained in terms of different competition between ionic species for the buffer and/or the glass electrode surface [100]. At the very least [100] shows that the same NES forces responsible for interfacial tensions and specific ion effects in proteins predict changes in pH of the right magnitude and cannot therefore be neglected.

4.7.1 A Comment on Junction Potential

Glass electrode functioning is based on a glass membrane that is sensible to H^+ activity. When calibrated against standard buffer solutions of known pH ($p\mathrm{H}_s$), in principle, it shows a Nernstian behavior (see Fig. 20). It is known that [71]:

$$pH_{measured} = pH_s - \frac{(E_s - E') + (E'_{LJ} - E_{LJ})}{2.303\,RT/F} \tag{11}$$

where, E_s is the potential of the standard buffer solution, E' is the potential of the unknown solution, E'_{LJ} and E_{LJ} are the liquid junction potentials of the unknown and the standard buffer solutions, respectively. Liquid junction potentials are caused by the different distribution of cations and anions at the interface between two different electrolyte solutions. This asymmetry is usually ascribed to different activities and diffusion velocities across the interface between the two solutions [77, 78, 101].

In a typical pH measurement, the inner solution of a glass electrode is always the same, and the difference E_{LJ} is due to the different composition between the standard and the unknown solution. Differences in E_{LJ} are usually ascribed to differences in ionic strength [71, 73] or to H^+ activity coefficient differences in the presence of different background salts (bulk effect) [76].

In view of old and recent findings related to Hofmeister effects [1] rationalized in terms of dispersion forces, an alternative explanation can be proposed. The potential E'_{LJ} changes with type and concentration of the electrolyte in the unknown solution; this is caused by ion–glass membrane interaction (surface effect), driven by the combination of ES and NES forces, which, we note, are not additive [52]. This will lead to ion specific $(E'_{LJ} - E_{LJ})$ differences and thus to ion specific $p\mathrm{H}_{measured}$ [78].

In a more sophisticated electrochemical treatment, the different contributions to the glass electrode response should be considered. Namely, the potential difference on the membrane, the ohmic drop, the internal and external reference electrode potentials, and the liquid junction potential arising from the migration of ions between the sample and the reference solution must be taken into account; the latter being probably the most affected by the variation of the background electrolyte composition and concentration [102]. However, in the present study, we report and discuss the effect of salt solutions on the overall measured pH of a buffer solution, and do not make any attempt at separating the individual outcome on each single potential.

5 Concluding Remarks

We have shown that conventional theories are not able to predict experimental measurements on pH of buffer solutions with increasing salt concentration. A first attempt to rationalize these effects was done by Boström et al. [100]. This analysis was incorrect since the experiments did not take into account CO_2 dissolution [103]. Nevertheless, the idea that dispersion forces acting on ions were sufficiently large to accommodate observed trends there and in buffers seemed to be interesting. Indeed, the dispersion forces contribution [52] has explained a number of Hofmeister phenomena, including interfacial tensions If changes in bulk solution properties determine these effects, one is left with the conundrum that specific ion adsorption alone also does. NES forces certainly exist and have not been taken into account properly.

However, the debate on ascribing Hofmeister effects to bulk or surface effects is, in fact, largely semantic. A more subtle bulk effect can be identified. Thus, the interactions of an ion with water, in the presence of its neighbors, involves contributions from the Born (electrostatic interactions) plus quantifiable specific NES many-body contributions that are included in the terms hydration [64]. The consequently "dressed" ions, with hydration "shells" that are soft or hard, kosmotropic or chaotropic depending on one's taste in terminology, can interact and overlap, more or less via combined ES and NES forces, to give rise to ion specific activities. The specificity seems to occur mainly via ionic dispersion forces that, within the same (Lifshitz) formalism, give both self and interaction energies. The same route can be accomplished, in principle, via molecular simulations. But it is somewhat simpler to use measured dielectric properties of solutions as a function of frequency to deduce the required energies. The same processes go on at any interface.

Our results seem to indicate that the adsorption of ions at the glass electrode/water interface is a non-negligible phenomenon. This conclusion is consistent with other interfacial phenomena [9, 30, 99], and with some more recent papers that show that only a few layers of hydrating solvent molecules are strongly affected by the ionic electric field, while the structure of bulk water remains unperturbed [39–41, 104]. So, a contribution that involves specific surface adsorption of ions at interfaces seems to be more relevant.

The wider implications we believe deserve mention. For, if what we believe is an experimental pH measurement relies on an interpretation based on incorrect assumptions, then the implications for extrapolating and using such measurements to infer surface pH and potential of a biomembrane, and the standard $pK_a s$ of proteins and colloids, is also an open question. These effects are important if one considers that the glass electrode procedure is the most common way for measuring pH, particularly in biological systems.

In the end, this chapter shows how the addition of different electrolytes affects the pH of a buffer solution measured with a glass electrode. This effect – which depends on the nature of salt and buffer solutions and on their concentration – is real and cannot be neglected. Qualitatively, the measured pH value may be affected by the presence of the electrolytes via bulk and surface interactions according to the scheme reported in Fig. 19. It would take us too far afield to consider yet another matter that is beginning to loom large. Dissolved atmospheric gas affects hydrophobic association and interactions and probably free radical production profoundly. Dissolved gas depends on salt concentration also dropping to essentially zero at 1 M concentration. This will probably open up an entirely new ball game in the future.

NES Forces to introduce ion specificity

$$U_{dispersion}(x) \approx \frac{(n_{water}^2 - n_{glass}^2)\alpha_i^*(0)\hbar\omega_i}{8x^3}$$

This ionic dispersion potential (NES) should be included in calculating ion real concentrations, particularly [H⁺] at the electrode surface

$$c_{\pm}(x) = c^0 \exp\left[-\frac{1}{k_B T}\left(\pm e\varphi + U_{dispersion}(x)\right)\right] = c^0 \exp\left[-\frac{\Phi}{k_B T}\right]$$

$$[H^+]_{surf} = [H^+]_{bulk} \exp\left[-\frac{\Phi}{k_B T}\right]$$

Fig. 20 The calculation of $[H^+]$ concentration. NES dispersion potential felt by an ion at the distance x from a surface depends on: (i) The refractive index of the substrate $n_{substrate}$ compared to that of water (ii) The static excess polarizability of the ion in water a_i (iii) The electron affinity (or ionization potential) $\hbar\omega_i$ is not known experimentally with any real certainty, but it should be between IR and UV frequency [64–66]

Ion specificity needs to be accounted for during pH measurements through suitable dispersion potentials according, for instance, to the equations reported in Fig. 20.

Enzymatic activity is clearly affected by buffers, added salts, and pH. Hofmeister effects appear as a natural consequence. The origin of the observed phenomena is related to both bulk and surface interaction. In turn, these interactions arise from the special properties of water molecules (solvation ability, strength of the hydrogen bond) in relation to the specific properties (particularly polarizability) of the added ions – anions and cations – that can show their features under different names: Kosmotropic and chaotropic ions, alias hard and soft ions, alias structure making and structure breaking ions, alias lyotropic series, alias Hofmeister series (cfr. Fig. 2).

Acknowledgment Fondazione Banco di Sardegna, CSGI (the Italian Consortium for colloid science and Nanotechnology), Italian MUR are acknowledged for financial support.

References

1. Kunz, W.; Nostro, P. L.; Ninham, B. W., *Curr. Op. Colloid Int. Sci.*, **9**, 1–18 (2004).
2. Robinson, R. A.; Stokes, R. H. *Electrolyte Solutions*; Butterworths: London, 1959.
3. Hofmeister, F., *Arch. Exp. Pathol. Pharmakol.*, **24**, 247–260 (1888).
4. Kunz, W.; Henle, J.; Ninham, B. W., *Curr. Op. Colloid Int. Sci.*, **9**, 19–37 (2004).
5. Loeb, J., *Science*, **52**, 449–456 (1920).
6. Gustavson, K. H. Specific ion effects in the behaviour of tanning agents toward collagen treated with neutral salts. In *Colloid Symposium Monograph*; Weiser, H. B., Ed.; The Chemical Catalog Company Inc.: New York, 1926.
7. Riés-Kautt, M. M.; Ducruix, A. F., *J. Biol. Chem.*, **264**, 745–748 (1989).
8. Carbonnaux, C.; Riés-Kautt, M. M.; Ducruix, A. F., *Protein Sci.*, **5**, 2123–2128 (1995).
9. Boström, M.; Tavares, F. W.; Finet, S.; Skouri-Panet, F.; Tardieu, A.; Ninham, B. W., *Biophys. Chem.*, **117**, 217–224 (2005).
10. Warren, J. C.; Cheatum, S. G., *Biochemistry*, **5**, 1702–1707 (1966).
11. Warren, J. C.; Stowring, L.; Morales, M. F., *J. Biol. Chem.*, **241**, 309–316 (1966).
12. Wondrak, E. M.; Louis, J. M.; Oroszlan, S., *FEBS Lett.*, **280**, 344–346 (1991).
13. Park, C.; Raines, R. T., *J. Am. Chem. Soc.*, **123**, 11472–11479 (2001).
14. Kim, H.-K.; Tuite, E.; Nordén, B.; Ninham, B. W., *Eur. Phys. J. E*, **4**, 411–417 (2001).
15. Wright, D. B.; Banks, D. D.; Lohman, J. R.; Hilsenbeck, J. L.; Gloss, L. M., *J. Mol. Biol.*, **323**, 327–344 (2002).
16. Ramos, C. H. I.; Baldwin, R. L., *Protein Sc.*, **11**, 1771–1778 (2002).
17. Terland, O.; Flatmark, T., *Biochem. J.*, **369**, 675–679 (2003).
18. Zoldàk, G.; Sprinzl, M.; Sedlàk, E., *Eur. J. Biochem.*, **271**, 48–57 (2004).
19. Pinna, M. C.; Salis, A.; Monduzzi, M.; Ninham, B. W., *J. Phys. Chem. B*, **109**, 5406–5408 (2005).
20. Pinna, M. C.; Bauduin, P.; Touraud, D.; Monduzzi, M.; Ninham, B. W.; Kunz, W., *J. Phys. Chem. B*, **109**, 16511–16514 (2005).
21. Low, P. S.; Somero, G. N., *Proc. Natl. Acad. Sci.*, **72**, 3305–3309 (1975).
22. Pahlich, E.; Gelleri, B.; Kindt, R., *Planta*, **138**, 161–165 (1978).
23. Craine, J. E.; Daniels, G. H.; Kaufman, S., *J. Biol. Chem.*, **248**, 7838–7844 (1973).
24. Kunz, W.; Belloni, L.; Bernard, O.; Ninham, B. W., *J. Phys. Chem. B*, **108**, 2398–2404 (2004).
25. Maheshwari, R.; Sreeram, K. J.; Dhathathreyan, A., *Chem. Phys. Lett.*, **375**, 157–161 (2003).
26. Gurau, M. C.; Lim, S.-M.; Castellana, E. T.; Albertorio, F.; Kataoka, S.; Cremer, P. S., *J. Am. Chem. Soc.*, **126**, 10522–10523 (2004).
27. Lo Nostro, P.; Fratoni, L.; Ninham, B. W.; Baglioni, P., *Biomacromolecules*, **3**, 1217–1224 (2002).
28. Lo Nostro, P.; Lopes, J. R.; Ninham, B.W.; Baglioni, P., *J. Phys. Chem. B*, **106**, 2166–2174 (2002).
29. Lo Nostro, P.; Ninham, B. W.; Ambrosi, M.; Fratoni, L.; Palma, S.; Allemandi, D.; Baglioni, P., *Langmuir*, **19**, 9583–9591 (2003).
30. Lonetti, B.; Lo Nostro, P.; Ninham, B. W.; Baglioni, P., *Langmuir*, **21**, 2242–2249 (2005).
31. Washabaugh, M. W.; Collins, K. D., *J. Biol. Chem.*, **261**, 12477–12485 (1986).
32. Collins, K. D., *Q. Rev. Biophys.*, **18**, 323–422 (1985).
33. Collins, K. D., *Proc. Natl. Acad. Sci.*, **92**, 5553–5557 (1995).
34. Collins, K. D., *Biophys. J.*, **72**, 65–76 (1997).
35. Hribar, B.; Southall, N. T.; Vlachy, V.; Dill, K. A., *J. Am. Chem. Soc.*, **124**, 12302–12311 (2002).

36. Romsted, L. S., *Langmuir*, **23**, 414–424 (2007).
37. Ninham, B. W., *Adv. Colloid Interface Sci.*, **1**, 1–17 (1999).
38. Salis, A.; Pinna, M. C.; Bilanicova, D.; Monduzzi, M.; Lo Nostro, P.; Ninham, B. W., *J. Phys. Chem. B*, **110**, 2949–2956 (2006).
39. Omta, A. W.; Kropman, M. F.; Woutersen, S., *J. Chem. Phys.*, **119**, 12457–12461 (2003).
40. Omta, A. W.; Kropman, M. F.; Woutersen, S.; Bakker, H. J., *Science*, **301**, 347–349 (2004).
41. Zhang, Y.; Furyk, S.; Bergbreiter, D. E.; Cremer, P. S., *J. Am. Chem. Soc.*, **127**, 14505–14510 (2005).
42. Voinescu, A. E.; Bauduin, P.; Pinna, M. C.; Touraud, D.; Ninham, B. W.; Kunz, W., *J. Phys. Chem. B*, **110**, 8870–8876 (2006).
43. Boström, M.; Ninham, B.W., *Colloids and Surfaces A-Physicochemical and Engineering Aspects*, **291**, 24–29 (2006).
44. Voet, D.; Voet, J. G. *Biochemistry*; J. Wiley & Sons, 1990.
45. Chaplin, M.; Bucke, C. Effect of pH and ionic strength. In *Enzyme Technology*; Cambridge University Press, 1990.
46. Boström, M.; Williams, D. R. M.; Ninham, B. W., *Curr. Op. Colloid Int. Sci.*, **9**, 48–52 (2004).
47. Goldstein, L.; Levin, Y.; Katchalski, E., *Biochemistry*, **3**, 1913 (1964).
48. Grochulski, P.; Li, Y.; Schrag, J. D.; Bouthillier, F.; Smith, P.; Harrison, D.; Rubin, B.; Cygler, M., *J.Biol.Chem.*, **268**, 12843–12847 (1993).
49. Grochulski, P.; Li, Y.; Schrag, J. D.; Cygler, M., *Protein Sci.*, **3**, 82–91 (1994).
50. Lee, K. K.; Fitch, C. A.; Lecomte, J. T. J.; Carcia-Moreno, E. B., *Biochemistry*, **41**, 5656–5667 (2002).
51. Boström, M.; Williams, D. R.; Ninham, B. W., *Biophys. J.*, **85**, 686–694 (2003).
52. Ninham, B. W.; Yaminsky, V., *Langmuir*, **13**, 2097–2108 (1997).
53. Ninham, B. W.; Bostrom, M., *Cellul. Molec. Biol.*, **51**, 803–813 (2005).
54. Ninham, B. W., *Prog. Colloid Polym. Sci.*, **120**, 1–12 (2002).
55. Derjaguin, B. V.; Landau, L., *Acta Physicochim. USSR*, **14**, 633 (1941)
56. Verwey, J.; Overbeek, J. T. G. *Theory of the stability of lyophobic colloids*; Elsevier: Amsterdam, 1948.
57. Lo Nostro, P.; Nostro, A. L.; Ninham, B. W.; Pasavento, G.; Fratoni, L.; Baglioni, P., *Curr. Op. Colloid Int. Sci.*, **9**, 97–101 (2004).
58. Nyden, M.; Soderman, O., *Langmuir*, **11**, 1537–1545 (1995).
59. Nyden, M.; Soderman, O.; Hansson, P., *Langmuir*, **17**, 6794–6803 (2001).
60. Karaman, M. E.; Ninham, B. W.; Pashley, R. M., *J. Phys. Chem.*, 11512–11518 (1994).
61. Alfridsson, M.; Ninham, B. W.; Wall, S., *Langmuir*, **16**, 10087–10091 (2000).
62. Manciu, M.; Ruckenstein, E., *Adv. Colloid Interface Sci.*, **105**, 63–101 (2003).
63. Ruckenstein, E.; Manciu, M., *Adv. Colloid Interface Sci.*, **105**, 177–200 (2003).
64. Boström, M.; Ninham, B. W., *J. Phys. Chem. B*, **108**, 12593-12595 (2004).
65. Boström, M.; Ninham, B. W., *Langmuir*, **20**, 7569–7574 (2004).
66. Boström, M.; Ninham, B. W., *Biophys. Chem.*, **114**, 95–101 (2005).
67. Blum, F. D.; Pickup, S.; Ninham, B. W.; Chen, S. J.; Evans, D. F., *J. Phys. Chem.*, **89**, 711 (1985).
68. Chen, S. J.; Evans, D. F.; Ninham, B. W.; Mitchell, D. J.; Blum, F. D.; Pickup, S., *J. Phys. Chem.*, **90**, 842–847 (1986).
69. Monduzzi, M.; Caboi, F.; Larché, F.; Olsson, U., *Langmuir*, **13**, 2184–2190 (1997).
70. Chen, V.; Evans, D. F.; Ninham, B. W., *J. Phys. Chem.*, **91**, 1823–1826 (1987).
71. Hedwig, G. R.; Powell, H. K. J., *Analytical Chemistry*, **43**, 1206–1212 (1971).
72. Garcia-Mira, M. M.; Sanchez-Ruiz, J. M., *Biophys. J.*, **81**, 3489–3502 (2001).
73. Brown, R. J. C.; Milton, M. J. T., *Accred Qual Assur*, **8**, 505–510 (2003).
74. Pooler, P. M.; M. L. Wahl; Rabinowitz, A. B.; Owen, C. S., *Analytical Biochem.*, **256**, 240–242 (1998).
75. Brandariz, I.; Barriada, J. L.; Taboada-Pan, C.; Sastre de Vicente, M. E., *Electroanalysis*, **13**, 1110–1114 (2001).

76. Brandariz, I.; T.Vilarino; P.Alonso; Herrero, R.; Fiol, S.; Sastre de Vicente, M. E., *Talanta*, **46**, 1469–1477 (1998).
77. Buck, R. P.; Rondinini, S.; Covington, A. K.; Baucke, F. G. K.; Brett, C. M. A.; Camoes, M. F.; Milton, M. J. T.; Mussini, T.; Naumann, R.; Pratt, K. W.; Spitzer, P.; Wilson, G. S., *Pure Appl. Chem.*, **74**, 2169–2200 (2002).
78. Schneider, A. C.; Pasel, C.; Luckas, M.; Schmidt, G. K.; Herbell, J.-D., *J. Sol. Chem.*, **33**, 257–273 (2004).
79. Kékicheff, P.; Ninham, B. W., *Europhysics Letters*, **12**, 471–477 (1990).
80. Nylander, T.; Kékicheff, P.; Ninham, B. W., *J. Colloid Interf. Sci.*, **164**, 136–150 (1994).
81. Pashley, R. M.; Ninham, B. W., *J. Phys. Chem.*, **91**, 2902–2904 (1987).
82. Finet, S.; Skouri-Panet, F.; Casselyn, M.; F.Bonnete; A.Tardieu, *Curr. Op. Colloid Int. Sci.*, **9**, 112–116 (2004).
83. Lo Nostro, P.; Ninham, B. W.; Lo Nostro, A.; Pesavento, G.; Fratoni, L.; Baglioni, P., *Pysical Biology*, **2**, 1–7 (2005).
84. Pinna, M. C.; Salis, A.; Monduzzi, M.; Ninham, B. W., *J. Phys. Chem. B*, **109**, 14752–14754 (2005).
85. Millero, F. J. Compilation of the partial molal volumes of electrolytes at infinite diluition, and the apparent molal volume concentration dependence constants, at various temperatures. In *Water and Aqueous Solutions*; Horne, R. A., Ed.; Wiley-Interscience: New York, 1972; pp 565–595.
86. Butler, J. N.; Cogley, D. R. *Ionic Equlibrium: Solubility and pH Calculations*; John Wiley & Sons: New York, 1998.
87. Kielland, J., *J. Am. Chem. Soc.*, **59**, 1675 (1937).
88. Jano, I.; Jarvis, T., *J. Sol. Chem.*, **31**, 317–339 (2002).
89. Fazary, A. E., *J. Chem. Eng. Data*, **50**, 888–895 (2005).
90. Partanen, J. I., *Acta Chem. Scand.*, **50**, 492–498 (1996).
91. Stern, O., *Elektrochem.*, **30**, 508 (1924).
92. Parsons, R. In *Modern Aspects of Electrochemistry*; Bockris, J. O. M., Ed.; Butterworths: London, 1954.
93. Grahame, D. C., *Chem. Rev.*, **47**, 441 (1947).
94. Lorenz, W.; Salié Zeit, G., *Phys. Chem.*, **218**, 259 (1961).
95. Schultze, J. W.; Koppitz, F. D., *Electrochim. Acta*, **21**, 327 (1976).
96. Conway, B. E., *Electrochimica Acta*, **40**, 1501–1512 (1995).
97. Conway, B. E.; Ayranci, E., *J. Solution Chem.*, **28**, 163–192 (1999).
98. Jarvis, N. J.; Schieman, M. A., *J. Phys. Chem.*, **72**, 74 (1968).
99. Boström, M.; Kunz, W.; Ninham, B. W., *Langmuir*, **21**, 2619–2623 (2005).
100. Boström, M.; Craig, V. S. J.; Albion, R.; Williams, D. R. M.; Ninham, B. W., *J. Phys. Chem. B*, **107**, 2875–2878 (2003).
101. Atkins, P. W. *Physical Chemistry*, Third edition ed.; Oxford University Press: Oxford, 1986.
102. Tishchenko, P. Y.; Chichkin, R. V.; Il¡ENT FONT=(normal text) VALUE=39¿'¡/ENT¿ina, E. M.; Wong, C. S., *Oceanology*, **42**, 32–41 (2002).
103. Westcott, C. C. *pH measurements*; Academic Press: Orlando, 1978.
104. Näslund, L. Å.; Edwards, D. C.; Wernet, P.; Bergmann, U.; Ogasawara, H.; Petterson, L. G. M.; Myneni, S.; Nilsson, A., *J. Phys. Chem. A*, **109**, 5995–6002 (2005).

Index

Actin fibers, 97
Adhesins, 73, 74
AFM, 91–92, 114
Agglomeration/dispersal, 2, 3, 16, 42, 64
Algorithm, 103
Antibacterial, 54, 116
Antibodies, 2, 3, 28, 53, 83, 92–93, 110–112, 114, 115, 116, 117
Antibody microarray, 82, 83

Bacterial colonization, 69, 72, 76, 98
Bioaccumulation, 2, 5, 19, 27, 28, 30
Bioadhesion, 69
Bioaffinity sensors, 70, 109–110, 111
Biochip, 53, 78, 81, 82
Biocompatibility, 1–43, 69, 74, 75, 77, 81, 89, 115
Biofilm, 72, 73, 98, 99, 103, 104
Biofilter, 70, 108–109, 116
Bio-functionalization, 70, 88, 109–118
Biological cell and tissue, 1, 3, 5, 15, 27, 29, 41, 42
Biological and medical applications, 1, 2, 47, 48, 62, 86, 72, 75, 111, 112
Biological surface modification, 90
Biomaterials, 2, 69, 70, 72, 74, 75, 77, 88, 90, 98, 100, 113
BioMEMS, 77–78, 79, 87, 104
Biomotor molecules, 86
Biopanning, 110
Bioreceptors, 110
Bioselective sensors, 110
Biosensors, 5, 53, 56, 57, 60, 62, 65, 69, 70, 72, 74, 75, 76, 78–80, 82, 86, 90, 94, 100, 101, 104, 105, 110, 111, 113, 116
Biotechnological applications, 54, 69, 74–75, 89, 104

Cantilever microarrays, 84
Carbon nanotubes, 3, 4, 6, 7, 12, 13, 14, 15, 17, 18, 47–65, 85, 86, 115–116
CD type platform, 81
Cell-based assays, 80
Cell-based sensor, 79, 90
Cell-free assays, 80
Cell/surface interactions, 5, 95, 176, 184, 187, 188–189
Cellular automata, 102
Cellular enzyme degradation, 19
Cellular interactions, 3
Cellular membranes, 5, 8, 18, 33, 39, 41, 42, 62
CEM, 105
Chemical compound microarray, 82, 83
CNNs, 102
CNTs, 6, 14, 16, 48, 54, 55, 56, 57, 58, 60, 61, 62, 64, 65, 86, 116
Co-adhesion, 72, 74
Computational model, 101, 103
Contact angle, 92
Contact guidance, 95, 97
Controlled cell adhesion, 69, 75–87, 98
Cytoskeleton, 71, 92, 97
Cytotoxicity, 2, 4, 8, 10, 12, 16, 17, 19, 21, 22, 25, 26, 28, 29, 30, 31, 32, 36, 80
Degradation and excretion, 2, 5, 19, 26, 30, 36, 41, 72, 76–77, 118
Delivery vehicles, 113, 116, 117
Dewetting, 100–101, 104
Diagnostics, 52–54, 65, 77, 78, 79, 81, 85
Diagnostics and tumor therapy, 48, 65
DLVO theory, 73–74, 175
DNA microarrays, 82, 83
DNA vaccines, 116, 117
Dose, 2, 3–4, 5, 10, 16–17, 21–22, 25, 26, 27, 29, 31, 32, 34, 42, 48, 87
Drug delivery, 2, 52–54, 65, 74, 77, 78, 113, 115

ECM, 3, 4, 29, 70, 71, 91, 92
Electrochemically active bacteria, 106
Electronic sensors, 115
Engineered nanoparticles, 1–2, 3, 4, 5, 27, 30
Environmental bioremediation, 107
EPS, 72, 73
Exposure time, 3, 5, 10, 13, 16
Extended DLVO theory, 73, 74

FN, 90
Focal contacts, 71, 93, 97
Fractal structure, 102, 104
Functionalization, 3, 29, 42, 55, 70, 85, 86, 88, 90, 91, 109–118

Gene delivery, 54, 118
Generic modelling, 103
Guided cell alignment, 95

H_2 production, 107
HTS, 80
Human lung and colon epithelial monolayers, 35

Immuno-fluorescence, 92, 93
In silico methods, 81
In-situ bioremediation, 109
Integrins, 70, 71, 90, 97
Irradiation treatments, 88, 90

Lab-on-a-chip, 81–82
Light and electron microscopy, 8, 30, 31
Living cells and tissues, 2, 29, 31, 41, 42, 43, 53, 56, 62,

M13, 110, 112, 113, 114
Mediators, 16, 86, 105, 106
MFC, 105, 105, 107, 108
μMFCs, 107, 108
Microarrays, 69, 82–84, 85, 112, 115
Microbial arrays, 70, 104, 105–109
Microbial filter, 105, 109
Micro-patterned surfaces, 70, 94, 96
Micropatterning, 90, 101, 104, 105, 108
MicroPET, 20, 21, 23, 24, 26, 43
Micro-TAS, 78
Mimetopes, 117
Mitotic spindles, 94
Molecular devices and machines, 131–156
Multiple fluorescent method, 95
MWNT, 86

Nano-fabrication, 89
Nanoparticle processing, 4, 32, 40, 41
Nanoparticles for biomedical use, 46, 50, 67, 72, 75, 110, 112
Nanoscale elevator, 87
Nanostructured composite materials, 47, 65
Nanotechnology, 47, 48–54, 69, 75, 77, 84–86, 114, 131, 155
Nano-wires, 86
NEMS, 86
NHDF cells, 95, 96, 97

Opto-electronic nanomaterials, 114

P. aeruginosa biofilm, 99
Patterning, 69, 72, 88, 89, 90, 100–109
PEC, 106
μPECs, 107, 108
Peptide chips, 83
Peptides, 26, 62, 69, 70, 74, 83, 90, 91, 109–118
Phage display technology, 91, 109, 111, 114
Phage probes, 111
Phages applications, 11
Phage therapy, 116, 117
Phenazine, 106
PIII, 110, 112, 113, 114
Plasma-surface modification, 89
Polymers, 69, 74, 81, 87, 89, 92, 111, 113
Porous materials, 155
Protein chips, 83
Protein microarrays, 82
PVIII, 110, 113, 114

Recombinant peptides, 110
Responses in vitro, 30–31
RGD, 90

Scaffold, 74, 76, 77, 88
Self-organization, 100, 101
Self-organized criticality, 102
SFE, 91
Signal peptides, 117
Soft lithography, 83
Spatial confinement, 69, 70, 87, 95, 100
Spatial control, 69, 75, 101
Spatial controlled adhesion, 69, 75
Surface engineering, 75, 79
Surface modification, 69, 76, 77, 87, 88, 89, 90, 91
Surface physico-chemical functionalization, 88
Surface reactivity, 3
Synchronized cells, 104
Synthetic materials, 74, 90

Targeted therapy, 118
μTAS, 78
Thermodynamic approach, 73, 74
Tissue engineering, 69, 74, 75, 76–77, 89, 91, 98
Tissue microarrays, 82, 83, 85
Toxicity of carbon nanotubes, 62–63

U937 cells, 96

Vaccine delivery vehicles, 117
VERO cells, 94, 95, 98

Water purification, 85
Whole-cell phage display, 111

XPS, 91

Printing: Krips bv, Meppel, The Netherlands
Binding: Stürtz, Würzburg, Germany